BURLEIGH DODDS SCIENCE: INSTANT INSIGHTS

NUMBER 96

# Optimising agri-food supply chains

AF787456

burleigh dodds
SCIENCE PUBLISHING

Published by Burleigh Dodds Science Publishing Limited
82 High Street, Sawston, Cambridge CB22 3HJ, UK
www.bdspublishing.com

Burleigh Dodds Science Publishing, 1518 Walnut Street, Suite 900, Philadelphia, PA 19102-3406, USA

First published 2024 by Burleigh Dodds Science Publishing Limited


British Library Cataloguing in Publication Data
A catalogue record for this book is available from the British Library

ISBN 978-1-80146-665-3 (Print)
ISBN 978-1-80146-666-0 (ePub)

DOI: 10.19103/9781801466660

Typeset by Deanta Global Publishing Services, Dublin, Ireland

# Contents

Acknowledgements ix

1 Studying agri-food supply chains: an analytical framework 1
*Sander de Leeuw, Renzo Akkerman and Rodrigo Romero-Silva, Wageningen University, The Netherlands*

1 Introduction 1
2 Agri-food supply chains: a brief review 2
3 The development of the concept of 'supply chains' 4
4 Food systems versus food supply chains 5
5 Designing agri-food supply chains: a framework 7
6 The structure of the book 7
7 References 8

2 Advances in traceability systems in agri-food supply chains 11
*Samantha Islam, University of Cambridge, UK; Louise Manning, Royal Agricultural University, UK; and Jonathan M. Cullen, University of Cambridge, UK*

1 Introduction 11
2 Traceability drivers 14
3 Food traceability in theory 16
4 Food traceability system 17
5 Traceability technologies 19
6 Barriers in implementation of effective food traceability systems 27
7 Improvement pathways 29
8 Where to look for further information 30
9 References 30

3 The role of technology in crisis management and product recall in food supply chains 37
*Louise Manning, Royal Agricultural University, UK; and Aleksandra Kowalska, Maria Curie-Skłodowska University, Poland*

1 Introduction 37

2 Food supply chain governance 39
3 Information asymmetry 41
4 Data sharing as a mechanism to underpin public-private partnerships (PPPs) 44
5 Concluding thoughts 48
6 Where to look for further information 49
7 References 50

4 Improving supply chains to prevent food losses and waste: an overview 59
*Hamid El Bilali, University of Natural Resources and Life Sciences (BOKU), Austria*

1 Introduction 59
2 Food losses and waste along the supply chain: main causes and reduction strategies 60
3 Reducing food wastage along the food supply chain: the need for a systems approach 67
4 Future trends and conclusion 71
5 Where to look for further information 72
6 References 72

5 Developing decision support systems for improving data management in agricultural supply chains 79
*Gerhard Schiefer, University of Bonn, Germany*

1 Introduction 79
2 Decisions in supporting data management 82
3 Decision tools 85
4 Principal case studies 88
5 Conclusion and future trends 93
6 References 94

# Series list

| Title | Series number |
| --- | --- |
| Sweetpotato | 01 |
| Fusarium in cereals | 02 |
| Vertical farming in horticulture | 03 |
| Nutraceuticals in fruit and vegetables | 04 |
| Climate change, insect pests and invasive species | 05 |
| Metabolic disorders in dairy cattle | 06 |
| Mastitis in dairy cattle | 07 |
| Heat stress in dairy cattle | 08 |
| African swine fever | 09 |
| Pesticide residues in agriculture | 10 |
| Fruit losses and waste | 11 |
| Improving crop nutrient use efficiency | 12 |
| Antibiotics in poultry production | 13 |
| Bone health in poultry | 14 |
| Feather-pecking in poultry | 15 |
| Environmental impact of livestock production | 16 |
| Sensor technologies in livestock monitoring | 17 |
| Improving piglet welfare | 18 |
| Crop biofortification | 19 |
| Crop rotations | 20 |
| Cover crops | 21 |
| Plant growth-promoting rhizobacteria | 22 |
| Arbuscular mycorrhizal fungi | 23 |
| Nematode pests in agriculture | 24 |
| Drought-resistant crops | 25 |
| Advances in detecting and forecasting crop pests and diseases | 26 |
| Mycotoxin detection and control | 27 |
| Mite pests in agriculture | 28 |
| Supporting cereal production in sub-Saharan Africa | 29 |
| Lameness in dairy cattle | 30 |
| Infertility and other reproductive disorders in dairy cattle | 31 |
| Alternatives to antibiotics in pig production | 32 |
| Integrated crop-livestock systems | 33 |
| Genetic modification of crops | 34 |

| | |
|---|---|
| Developing forestry products | 35 |
| Reducing antibiotic use in dairy production | 36 |
| Improving crop weed management | 37 |
| Improving crop disease management | 38 |
| Crops as livestock feed | 39 |
| Decision support systems in agriculture | 40 |
| Fertiliser use in agriculture | 41 |
| Life cycle assessment (LCA) of crops | 42 |
| Pre- and probiotics in poultry production | 43 |
| Poultry housing systems | 44 |
| Ensuring animal welfare during transport and slaughter | 45 |
| Conservation tillage in agriculture | 46 |
| Tropical forestry | 47 |
| Soil health indicators | 48 |
| Improving water management in crop cultivation | 49 |
| Fungal diseases of apples | 50 |
| Using crops as biofuel | 51 |
| Septoria tritici blotch in cereals | 52 |
| Biodiversity management practices | 53 |
| Soil erosion | 54 |
| Integrated weed management in cereal cultivation | 55 |
| Sustainable forest management | 56 |
| Restoring degraded forests | 57 |
| Developing immunity in pigs | 58 |
| Bacterial diseases affecting pigs | 59 |
| Viral diseases affecting pigs | 60 |
| Developing immunity in poultry | 61 |
| Managing bacterial diseases of poultry | 62 |
| Proximal sensors in agriculture | 63 |
| Dietary supplements in dairy cattle nutrition | 64 |
| Dietary supplements in poultry nutrition | 65 |
| Intercropping | 66 |
| Managing arthropod pests in tree fruit | 67 |
| Weed management in regenerative agriculture | 68 |
| Integrated pest management in cereal cultivation | 69 |
| Economics of key agricultural practices | 70 |
| Nutritional benefits of milk | 71 |

Biostimulant applications in agriculture 72
Phosphorus uptake and use in crops 73
Optimising pig nutrition 74
Nutritional and health benefits of beverage crops 75
Artificial Intelligence applications in agriculture 76
Ensuring the welfare of laying hens 77
Ecosystem services delivered by forests 78
Improving biosecurity in livestock production 79
Rust diseases of cereals 80
Optimising rootstock health 81
Irrigation management in horticultural production 82
Improving the welfare of gilts and sows 83
Improving the shelf life of horticultural produce 84
Improving the health and welfare of heifers and calves 85
Managing arthropod pests in cereals 86
Optimising photosynthesis in crops 87
Optimising quality attributes in poultry products 88
Advances in fertilisers and fertiliser technology 89
Sustainable tropical forest management 90
Phenotyping applications in agriculture 91
Fungicide resistance in cereals 92
Unmanned aircraft systems in agriculture 93
Using manure in soil management 94
Zero/no till cultivation 95
Optimising agri-food supply chains 96
Optimising quality attributes in horticultural products 97
Improving the sustainability of dairy production 98
Improving the welfare of growing and finishing pigs 99
Ensuring the welfare of broilers 100
Biofertiliser use in agriculture 101
Infertility and other reproductive disorders in pigs 102
Regenerative techniques to improve soil health 103
Agroforestry practices 104
Carbon monitoring and management in forests 105
Viruses affecting horticultural crops 106
Regulatory frameworks for new agricultural products and technologies 107
Machine vision applications in agriculture 108

| | |
|---|---|
| Alternative sources of protein for poultry | 109 |
| Alternative sources of protein for pigs | 110 |
| Nitrogen-use efficiency | 111 |
| Good agricultural practices (GAP) | 112 |

# Acknowledgements

Chapters in this Instant Insight are taken from the following sources:

Chapter 1 Introduction: types of agri-food supply chain

Chapter taken from: de Leeuw, S., Akkerman, R. and Romero-Silva, R. (ed.), Frontiers in agri-food supply chains: Frameworks and case studies, pp. 1-10, Burleigh Dodds Science Publishing, Cambridge, UK, 2024, (ISBN: 978 1 80146 271 6)

Chapter 2 Advances in traceability systems in agri-food supply chains

Chapter taken from: Manning, L. (ed.), Developing smart agri-food supply chains: Using technology to improve safety and quality, Burleigh Dodds Science Publishing, Cambridge, UK, 2022, (ISBN: 978 1 78676 749 3)

Chapter 3 The role of technology in crisis management and product recall in food supply chains

Chapter taken from: Manning, L. (ed.), Developing smart agri-food supply chains: Using technology to improve safety and quality, Burleigh Dodds Science Publishing, Cambridge, UK, 2022, (ISBN: 978 1 78676 749 3)

Chapter 4 Improving supply chains to prevent food losses and waste: an overview

Chapter taken from: Yahia, E. M. (ed.), Preventing food losses and waste to achieve food security and sustainability, Burleigh Dodds Science Publishing, Cambridge, UK, 2019, (ISBN: 978 1 78676 300 6)

Chapter 5 Developing decision support systems for improving data management in agricultural supply chains

Chapter taken from: Armstrong, L. (ed.), Improving data management and decision support systems in agriculture, Burleigh Dodds Science Publishing, Cambridge, UK, 2019, (ISBN: 978 1 78676 340 2)

# Chapter 1

# Studying agri-food supply chains: an analytical framework

*Sander de Leeuw, Renzo Akkerman and Rodrigo Romero-Silva, Wageningen University, The Netherlands*


1 Introduction
2 Agri-food supply chains: a brief review
3 The development of the concept of 'supply chains'
4 Food systems versus food supply chains
5 Designing agri-food supply chains: a framework
6 The structure of the book
7 References


## 1 Introduction

Food supply chains are an essential part of our society. From the primary production of agricultural products to the retail and foodservice settings through which food products are distributed, these extensive supply chains provide a wide range of products to satisfy daily demands for human nutrition. In addition, these food supply chains also provide a significant contribution to the economy and to the employment across the world (Beckman and Countryman, 2021).

We probably live in one of the most interesting times to study agri-food supply chains (AFSCs). Demand for food is expected to rise significantly over the next few decades, both as a result of population growth and even more so due to the rise in average income levels. Increases in agricultural food production (mostly due to improved yields) have so far been keeping up with increasing demand,[1] even though food insecurity remains a major problem in many parts of the developing and developed world. Matching food supply and demand is thus one of the key global challenges for the future.

1 https://ourworldindata.org/yields-vs-land-use-how-has-the-world-produced-enough-food-for-a-growing-population.https://hbr.org/2016/04/global-demand-for-food-is-rising-can-we-meet-it.

http://dx.doi.org/10.19103/AS.2023.0122.01

In addition to this challenge of rising demand, AFSCs regularly have to deal with external events that can lead to significant supply chain disturbances. Since the start of the COVID-19 pandemic around the turn of the year 2019–2020, for example, the media have almost continually reported problems with supply chains and resulting disruptions in food supplies. Examples include the closure of meat processing facilities due to COVID-19 outbreaks among staff (e.g. *Guardian*, 2020) or the occurrence of agricultural surpluses (with resulting losses and waste) due to mismatches in supply and demand and interruptions in supply chain operations (e.g. *Los Angeles Times*, 2020; *Washington Post*, 2020). The war in Ukraine has made us realize that the development of globalized agri-food chains has made many countries heavily dependent on agriculture in particular regions. The war has both disrupted production in Ukraine itself and closed the Black Sea, resulting in shortages in grain and consequently considerable price increases for both grain and commodities using grains such as bread.[2] Ukraine is also a major exporter of sunflower oil, which became a scarce commodity in many places soon after the start of the war.[3] The resulting price increases in sunflower oil have since then had an effect on pricing and availability of products that use sunflower oil as an ingredient such as snacks.

Partly due to these events, the design and development of effective global strategies for managing the supply chain of agri-food products has become an even more complex and challenging issue. An added dimension is ever-increasing consumer requirement (which now encompass not just price and quality but other issues such as environmental and social stability or animal welfare), which require responding to almost continuous changes in lifestyle and dietary preferences.

AFSCs have unique characteristics and challenges for which we need an appropriate analytical framework. In addressing these issues, we first discuss AFSC characteristics, after which we provide a framework for designing AFSCs.

## 2 Agri-food supply chains: a brief review

An AFSC comprises various stakeholders representing the different stages related to agricultural production, post-harvest, storage, processing, distribution and retail operations, and linkages between these components (Behzadi et al., 2018). The management and design of AFSCs has received considerable attention from both researchers and practitioners over recent decades. Although AFSCs have similarities with other manufacturing supply chains, they have specific characteristics that make the management of these chains more challenging than other manufacturing supply chains (Onggo

2 https://www.nbcnews.com/news/world/russia-ukraine-war-grain-blockade-global-food-crisis-rcna25910.
3 https://time.com/6155095/sunflower-oil-russia-ukraine/.

et al., 2019). Some examples are fluctuations in raw material supply (e.g. due to weather conditions which are becoming more unpredictable and extreme with climate change), food price volatility, food waste, food and nutrition security, and specific power and governance issues due to regulation frameworks governing food production (Ali et al., 2019). In addition, each of the AFSC stakeholders faces specific and unique challenges compared to other types of supply chains when it comes to running their operations in the supply chain (Mccullough et al., 2010). Farmers face issues with resource availability and fertilizer and pesticide use, amongst others (Yadav et al., 2022). Wholesalers and retailers face challenges such as variability in quality and quantity of farm-based inputs, seasonality in supply, and requirements for managing perishable food products in transport and storage, while shelf-life constraints and related raw material/finished product quality deterioration over time are an issue in all parts of the AFSC (van der Vorst, 2000).

The foremost issue that emerges both in practice and in the academic literature relates to food losses and waste (FLW). It is often argued that FLW accounts for one-third of total food produced, with the UN arguing that food waste is a particular problem at consumer level in developed countries in particular (UNEP, 2021). FLW is not only a problem because of loss of valuable food product but also because many people still do not have access to sufficient food or a balanced diet. Food insecurity has been on the rise in recent years, even in developed countries (FAO, 2021). A number of initiatives aim to tackle this, including food banks in countries such as the Netherlands which redirect surplus food to low-income families, simultaneously contributing to ameliorating food insecurity and FLW. They typically rely on donated food (e.g. provided free by retail outlets and consumers) and only rarely use financial donations to buy food.

FLW also represents a loss of resources, including water, inputs, energy and land (Scherhaufer et al., 2018). Water use for food production and processing is the biggest factor affecting global water scarcity (Weinzettel and Pfister, 2019). The FAO has established that 70% of freshwater withdrawals are for food production. In addition to the energy required in food production (either directly in for example fuel for farm machinery or indirectly for example in the energy required to manufacture inputs such as fertilizers and pesticides), several types of food (particularly fresh produce) are also often transported over long distances and may require refrigeration to stay fresh, all of which requires fuel or electricity. These resources could have been used more productively had FLW been prevented.

Closely related to the FLW issue is sustainability. It is well known that agriculture is a major contributor to greenhouse gas (GHG) emissions (Poore and Nemeck, 2018). In AFSCs the dominant share of emissions (around 80%) comes from farming, representing up to 24% of global GHG emissions

(Kucukvar and Samadi, 2015). However, it has been estimated that transport accounts for about 20% of food systems-related emissions (Li et al., 2022). In fact, one could argue that that FLW should be viewed from the perspective of sustainability to get a correct view of its impact: wasting 1 kg of meat has a much bigger impact on the environment than wasting 1 kg of vegetables, simply because of the higher level of inputs needed to produce meat (Van Rooijen et al., 2024). Reviews of AFSC reviews such as Yadav et al. (2022) mainly focus on the environmental and economic aspects of sustainability. The social implications of sustainability (such as effects on labour conditions or farmer behavior) are far less studied.

An often-quoted issue in AFSCs relates to food safety and food quality. In 2012, for example, Europe witnessed a major horsemeat scandal with research showing that consumers were particularly concerned about the authenticity (as well as safety) of meat products (Agnoli et al., 2016). Traceability of food has become a topic of concern that is often raised in research (cf. Yadav et al., 2022). However, food traceability systems are still relatively undeveloped compared to other supply chains (Feng et al., 2020). Blockchain technology, which is a distributed and decentralized system of time-stamped and linked blocks of information, is considered a promising technology to help increase transparency (Feng et al., 2020).

All these factors mean that there are many important and distinctive characteristics to be taken into account when designing AFSCs which, in other types of supply chain, would not play such an important role. These particularly relate to quality decay and perishability of food that should be accounted for in areas of supply chain design such as network modelling, distribution handling strategies, or inventory management. Several models have been developed to address different aspects of these challenges (see Yadav et al. (2021) for examples). The remainder of this chapter focuses on a design approach to address key challenges in AFSCs.

## 3 The development of the concept of 'supply chains'

An early reference to the notion of supply chains is by Professor Jay Forrester in his seminal book *Industrial Dynamics* (1961), which he referred to as supply pipelines. Many people point to the work of military theorists such as Sun Tzu (*The Art of War*) or Von Clausewitz (*On War*) on the challenges in supplying armies as the origin of thinking about supply chains. It is less well known that commercial organizations picked up the idea of building and managing relatively complex AFSCs many years ago. Amongst others, the East Indies Company based in Amsterdam (in the Netherlands) created their own supply network in the seventeenth and eighteenth centuries, building and operating a global shipping fleet to transport spices and other valuable food products from sources of production in Asia to consumer markets in Europe.[4] Between 1700 and 1730, the Zaanstreek area north of Amsterdam was already a logistics hub before the term had been invented. It was the shipbuilding heart of Europe with, at its height, a production output of almost one seagoing ship built every 2 days. Many principles that we see in supply chains were already applied back then. As an example, saw mills were situated close to the wharfs to avoid carrying heavy wooden components over long distances. More distant factories produced canvas for the sails or biscuits for food during the voyage since both were relatively light and could be transported easily over greater distances. In these more rural areas labour was also cheaper to keep costs down. Circular supply chain principles were also already applied: waste timber, for example, was used to build wharfs which also helped manage costs. In addition, a range of wood suppliers were used to ensure critical supplies kept coming in (cf. Daly et al., 2021).

Typically, most food supply chains in this period were still local or regional, with networks of local farms supplying neighbouring towns and cities. The development of railroads, advances in shipping (e.g. the development of steam power) and early refrigeration technologies in the nineteenth and early twentieth centuries stimulated the movement of food over longer distances (e.g. import of grain from North America and meat and dairy products from South America into Europe). However, it took until well into the twentieth century for supply chains to really explode globally. In particular, the development of the container made shipping over long distances more economical, making way for the development of regional and global third-party logistic networks (Notteboom and Rodrigue, 2006). It is now hard to imagine global food transport without the use of (in particular refrigerated) containers. Continual innovations in food packaging have also further expanded options for transporting food globally (see Risch, 2009, for an historical overview of developments in food packaging).

## 4 Food systems versus food supply chains

The notion of managing supply chains has long been acknowledged in food and agriculture research, albeit with more policy-oriented approach often focused on the concept of a 'food system'. The International Food Policy Research Institute (IFPRI) defines a food system as '...the sum of actors and interactions along the food value chain–from input supply and production of crops, livestock, fish, and other agricultural commodities to transportation, processing, retailing, wholesaling, and preparation of foods to consumption

4 This is further described in the canon of the Netherlands (https://www.canonvannederland.nl/en/noord-holland/zaanstreek/scheepsbouw).

and disposal'[5]. Even though this definition seems to have food supply chains at its core, food systems research does not usually focus on the practical issues in management of flows of goods from farm to consumer, which is where AFSC research can make a contribution.

There are two main ways in which a food systems perspective and a food supply chain perspective differ: scope and scale. Scope refers to the actors involved. Early supply chain research mostly considered the decision-making processes of commercial firms. The growing focus on sustainability has required a broader approach. In their seminal paper, Kleindorfer et al. (2005) call this the extended supply chain. The inclusion of environmental and social impacts in addition to classic economic considerations requires including the domains of government policy/regulation and societal expectations, meaning that actors such as governments, consumers, employees and non-governmental organizations (NGOs) also play a role.

This brings us to the second way the food systems perspective and the food supply chain perspective differs: scale. Typically, the food system perspective takes a more macro view as opposed to a micro view of businesses. This means that studies of food systems typically look at developments at a sector or product level, often including the biological systems that food production interacts with (e.g. by looking at areas such as how agriculture affects, or is affected by, the physical, chemical and biological health of the soils in which crops are grown). As an example, an important recent discussion is on the role of circular economy principles in the sustainability of food systems, in which different land use scenarios can be analysed on a national or even global level (see Van Zanten et al., 2023, for an analysis on the EU level). How this then translates to individual businesses and supply chains operating in the food system is typically not the focus of such studies. Clearly, both the food systems perspective and the food supply chain perspective are important contributors to the body of knowledge on the food sector.

Figure 1 provides a picture of a generalized AFSC, which consists of farming activities, processing activities (which relate to preparing the farm products for consumption where needed), distribution from farm and/or processing facilities to food service and retail outlets where consumers obtain food for consumption. Different companies in AFSCs cooperate strategically in one or more areas as legally independent and autonomous organizations. As a result, companies may be active in different AFSCs and play a different role in each. It is also important to note that AFSCs are not just focused on ensuring a safe, efficient and cost-effective flow of food products to the end consumer. They must also consider problems such as FLW which highlight broader resource management issues that must be addressed in AFSCs (Akkerman and Cruijssen, 2024).

5 See https://www.ifpri.org/topic/food-systems for more on food systems.

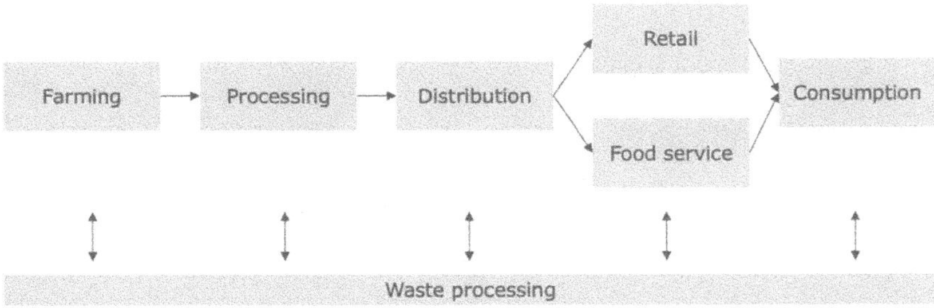


**Figure 1** Systems approach to managing supply chains. Source: Adapted from: van der Vorst (2000) and de Leeuw (1988).

## 5 Designing agri-food supply chains: a framework

Frameworks for the design and management of supply chains have been a topic of research for a few decades. Bemelmans (1982) was one of the first to develop a framework for information systems management in supply chains. Fundamental to his work was the need to develop processes and control in supply chains first, before starting to design information systems to manage them. Van der Vorst (2000) based his supply chain analysis framework on this concept and combined it with the systems approach of de Leeuw (1988). De Leeuw (1988) uses a systems approach and differentiates between the managed system, the managing system and the information system. Transformations take place in the managed system, which is where services or products are made. The information system is used to take stock of the relevant data that are generated during transformation. This results in key performance information or other types of information that is necessary to measure whether the output of the system meets the targets set. The managing system focuses on realizing this output by means of control variables (decision variables such as production volume), taking into account environmental factors (e.g. disruptions or illness). Each aspect influences key performance outcomes.

## 6 The structure of the book

This book is split into three parts. In Part 1, we first outline the state of development of AFSCs across the globe. In Chapter 2, Schoenherr et al. discuss the development of AFSCs in the Americas, after which Bourlakis et al. provide an overview of developments in northern and western Europe in Chapter 3. Ruiz et al. discuss the state of development and challenges in Southern and Eastern Europe in Chapter 4. Teng et al. close this section in Chapter 5 with an overview of developments in Asia.

In Part 2, following the structure of Fig. 2, we outline the challenges and opportunities in each of the parts. In Chapter 6, Aramyan et al. discuss the

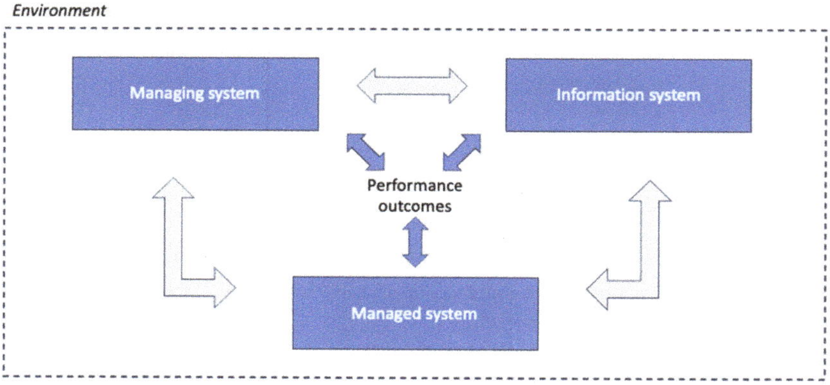


**Figure 2** Agri-food supply chain model.

measurement and management performance in AFSCs, after which Soysal et al. review the literature on AFSC infrastructure in Chapter 7. In Chapter 8, Transchel et al. discuss developments in planning and control of AFSCs, after which information system developments are discussed in Chapter 9 by Schiefer.

Part 3 is devoted to case studies on how to improve AFSCs across the globe. In Chapter 10, Soleiro et al. discuss the Americas, after which Akkerman et al. discuss developments in Europe based on a case study of the COVID-19 pandemic in Chapter 11. Jordaan et al. discuss opportunities in Africa in Chapter 12, and Li et al. do this for Asia in Chapter 13. Concluding remarks and future research opportunities in AFSCs are provided in Chapter 14.

## 7 References

Agnoli, L., Capitello, R., De Salvo, M., Longo, A. and Boeri, M. (2016). Food fraud and consumers' choices in the wake of the horsemeat scandal. *British Food Journal*. 118(8), 1898–1913.

Akkerman, R. and Cruijssen, F. (2024). Food loss, food waste, and sustainability in food supply chains. In: Bouchery, Y., Corbett, C. J., Fransoo, J. C. and Tan, T. (eds). *Sustainable Supply Chains: A Research-Based Textbook on Operations and Strategy* (2nd edn.). Springer, in press.

Ali, S. M., Moktadir, M. A., Kabir, G., Chakma, J., Rumi, M. J. U. and Islam, M. T. (2019). Framework for evaluating risks in food supply chain: implications in food wastage reduction. *J. Clean. Prod.* 228, 786–800. https://doi.org/10.1016/j.jclepro.2019.04.322.

Beckman, J. and Countryman, A. M. (2021). The importance of agriculture in the economy: impacts from COVID-19. *American Journal of Agricultural Economics* 103(5), 1595–1611.

Behzadi, G., O'Sullivan, M. J., Olsen, T. L. and Zhang, A. (2018). Agribusiness supply chain risk management: a review of quantitative decision models. *Omega* 79, 21–42. https://doi.org/10.1016/j.omega.2017.07.005.

Bemelmans, T. M. A. (1982). *Bestuurlijke informatiesystemen en automatisering*. Kluwer bedrijfswetenschappen (In Dutch).

Daly, A., Domínguez-Delmás, M. and van Duivenvoorde, W. (2021). Batavia shipwreck timbers reveal a key to Dutch success in 17th-century world trade. *PLoS ONE* 16(10), e0259391.

FAO (2021). Transforming food systems for food security, improved nutrition and affordable healthy diets for all. *The State of Food Security and Nutrition in the World*. Available at: http://www.fao.org/3/cb4474en/online/cb4474en.html#chapter-51.

Feng, H., Wang, X., Duan, Y., Zhang, J. and Zhang, X. (2020). Applying blockchain technology to improve agri-food traceability: a review of development methods, benefits and challenges. *Journal of Cleaner Production* 260, 121031.

Forrester, J. W. (1961). *Industrial Dynamics*. Productivity Press, Cambridge, MA.

Guardian (2020). 'Chaotic and crazy': meat plants around the world struggle with virus outbreaks. *The Guardian*, May 11, 2020.

Kleindorfer, P. R., Singhal, K. and van Wassenhove, L. N. (2005). Sustainable operations management. *Production and Operations Management* 14(4), 482–492.

Kucukvar, M. and Samadi, H. (2015). Linking national food production to global supply chain impacts for the energy-climate challenge: the cases of the EU-27 and Turkey. *Journal of Cleaner Production* 108, 395–408.

Li, M., Jia, N., Lenzen, M., Malik, A., Wei, L., Jin, Y. and Raubenheimer, D. (2022). Global food-miles account for nearly 20% of total food-systems emissions. *Nature Food* 3(6), 445–453. https://doi.org/10.1038/s43016-022-00531-w.

Los Angeles Times (2020). Rotting food. Hungry masses. Chaotic supply chains. Coronavirus upends the U.S. food system. *Los Angeles Times*, May 5, 2020.

McCullough, E., Stamoulis, K. and Pingali, P. (Eds) (2010). *The Transformation of Agri-Food Systems: Globalization, Supply Chains and Smallholder Farmers*. Earthscan, London.

Notteboom, T. and Rodrigue, J.-P. (2006). Containerisation, box logistics and global supply chains: the integration of ports and liner shipping networks. *International Journal of Maritime Economics* 10(1–2), 152–174.

Onggo, B. S., Panadero, J., Corlu, C. G. and Juan, A. A. (2019). Agri-food supply chains with stochastic demands: a multi-period inventory routing problem with perishable products. *Simulation Modelling Practice and Theory* 97, 101970. https://doi.org/10.1016/j.simpat.2019.101970.

Poore, J. and Nemecek, T. (2018). Reducing food's environmental impacts through producers and consumers. *Science* 360(6392), 987–992.

Risch, S. J. (2009). Food packaging history and innovations. *Journal of Agricultural and Food Chemistry* 57(18), 8089–8092.

Scherhaufer, S., Moates, G., Hartikainen, H., Waldron, K. and Obersteiner, G. (2018). Environmental impacts of food waste in Europe. *Waste Management* 77, 98–113.

UNEP (2021). *Food Waste Index Report 2021*. UNEP. Available at: https://www.unep.org/resources/report/unep-food-waste-index-report-2021.

van der Vorst, J. G. A. J. (2000). Effective food supply chains: generating, modelling and evaluating supply chain scenarios. Doctoral dissertation.

Van Rooijen, M. A., Gerdessen, J. C., Claassen, G. D. H. and de Leeuw, S. L. J. M. (2024). Optimizing household food waste: the impact of meal planning, package sizes, and performance indicators. Working paper, Wageningen University, Wageningen, the Netherlands.

van Zanten, H. H. E., Simon, W., van Selm, B., Wacker, J., Maindl, T. I., Frehner, A., Hijbeek, R., van Ittersum, M. K. and Herrero, M. (2023). Circularity in Europe strengthens the sustainability of the global food system. *Nature Food* 4(4), 320–330.

Washington Post (2020). Extra food is rotting on farms while Americans go hungry. *The Washington Post*, June 10, 2020.

Weinzettel, J. and Pfister, S. (2019). International trade of global scarce water use in agriculture: modeling on watershed level with monthly resolution. *Ecological Economics* 159, 301–311.

Yadav, V. S., Singh, A. R., Gunasekaran, A., Raut, R. D. and Narkhede, B. E. (2022). A systematic literature review of the agro-food supply chain: challenges, network design, and performance measurement perspectives. *Sustainable Production and Consumption* 29, 685–704.

# Chapter 2

# Advances in traceability systems in agri-food supply chains

*Samantha Islam, University of Cambridge, UK; Louise Manning, Royal Agricultural University, UK; and Jonathan M. Cullen, University of Cambridge, UK*

1 Introduction
2 Traceability drivers
3 Food traceability in theory
4 Food traceability system
5 Traceability technologies
6 Barriers in implementation of effective food traceability systems
7 Improvement pathways
8 Where to look for further information
9 References

## 1 Introduction

Food is a key building block of human physical well-being. Consumers' food purchasing behaviour is influenced by three key food characteristics: safety, quality and authenticity (Wilcock et al., 2004; Kendall et al., 2019). However, the repetitive occurrence of tragic and costly food crises, such as mad cow disease, dioxin contamination, the horse meat scandal, *Escherichia coli* (*E. coli*) outbreaks and *Salmonella* contamination not only diminishes consumers' confidence in food safety, quality and authenticity but also challenges the underlying credibility of the food industries.

The impact of such food safety incidents is amplified by the globalisation of the food trade, which causes transnational food safety issues (FAO, 2019). According to the World Health Organisation (WHO, 2015), approximately 600 million cases of global illness and 420 000 deaths are caused annually by foodborne diseases, and this leads to considerable economic burden owing to the associated use of healthcare services, product recalls and disposal and loss of sales and export. As estimated by the United States Department of Agriculture (USDA), the yearly cost of foodborne illnesses in the US stands at around \$10–83 billion (McLinden et al., 2014), while a single foodborne

http://dx.doi.org/10.19103/AS.2021.0097.01

pathogen, *E. coli*, is responsible for an expenditure of $14.42 million a year for healthcare use in the UK (Naylor et al., 2017). On the other hand, border rejections of export consignments result in significant economic loss; for example, owing to the detection of banned antibiotics and *Salmonella*, sea food supply chains in Asian lower- and middle-income countries are routinely disrupted (Blank, 2018). All these incidents, coupled with a more educated and aware public, underpin the increased demand for improved food traceability and for communication of information vis-à-vis food origins, ingredients, processing, quality and safety throughout food supply chains (FSCs) (Rodriguez-Salvador and Dopico, 2020).

Food traceability has been defined by many different organisations, in legislation and in scientific articles. The commonly used definition from ISO 22005 (ISO, 2005) refers to traceability as 'the ability to follow the movement of a feed or food through specified stage(s) of production, processing and distribution' (Fig. 1).

The implementation of food traceability requires the adoption of a food traceability system (FTS) – a specific arrangement of processes such as data capture, data storage and data exchange – that is capable of maintaining and communicating the desired product information through all stages of the FSC

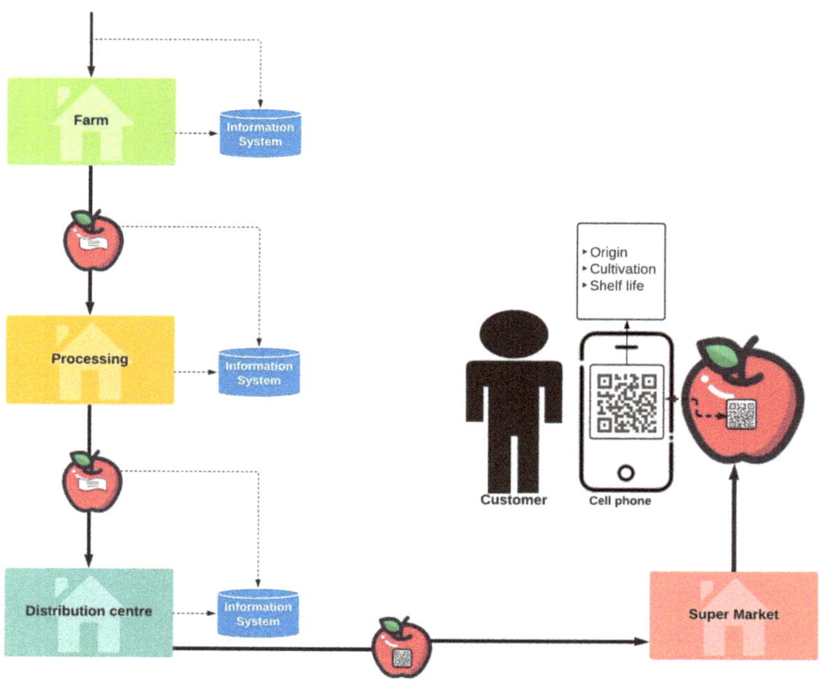


**Figure 1** Traceability system in the food supply chain.

(World Economic Forum, 2019). FSCs are becoming increasingly globalised, which poses challenges in guaranteeing symmetry, trust and transparency for the shared product information. Due to their complex and interconnected nature, global FSCs face difficulty in seeking to adopt a single, centralised traceability system architecture - especially if this is in the control of third parties that are monopolistic and opaque in nature. Further, such a centralised traceability system architecture is more vulnerable to collapse, since a single point of breakdown can cause the whole system to crash (Tian, 2017; El Maouchi et al., 2018). Distributed traceability system architectures have emerged as a response to these risks, making use of decentralised information capture and storage to deliver improved information symmetry, security, trust and transparency.

Although manual paper-based traceability systems are still commonly used, deployment of computerised FTSs is rising due to the rapid growth of information and communication technologies (ICTs) and the emergence of Industry 4.0 (the so-called fourth industrial revolution) marked by the automation of production systems. The advancement of FTSs is clearly noticeable in the identification technology practices employed in FSCs, such as the use of radio frequency identification (RFID) which provides increased data capacity, reading speeds and accuracy (McCathie and Michael, 2005; Badia-Melis et al., 2015) and is displacing the extensively used barcode technologies. Wireless sensors can be used to supplement RFID and support traceability-based product quality monitoring (Thakur and Forås, 2015; Alfian et al., 2020; Islam et al., 2021a). These technologies, when integrated with electronic product code information service (EPCIS) data standards, enable efficient product data transfer between supply chain partners (Mainetti et al., 2013) and through to customers. Moreover, blockchain and smart contract technologies allow increased opportunity for transparency and tamperproof data recording in complex and globally distributed supply chains (Pearson et al., 2019).

Despite the availability of such revolutionary technologies, there are various barriers inherent in food systems which impede successful implementation of FTSs in individual companies or in sub-sections of supply chains (Bosona and Gebresenbet, 2013; Hardt et al., 2017; World Economic Forum, 2019). The barriers to the adoption of FTSs include a lack of awareness and training on traceability and traceability technologies; a lack of standards development; resource deficiencies, including funding and capacity issues; inefficient data management; and finally, technology scaling issues (Islam and Cullen, 2021). The intensity of these barriers is influenced by the supply chain structure, its relationship dynamics, and the position of companies within the supply chain (Sterling et al., 2015). To harness the transformative power of breakthrough technologies and exploit the potential for FTSs to transform, FSCs will require clear strategies and pathways to overcome the barriers identified above.

This chapter considers exclusively the advancements in agri-food traceability systems and technologies, the barriers to their implementation and some potential improvement pathways to overcome these.

## 2 Traceability drivers

Implementation of FTSs is driven by several motivating factors, which are known as controlling forces (Norton et al., 2014) or *traceability drivers* in the literature. Traceability drivers differ across FTSs depending on what type of information the internal or external stakeholders require. A set of five prominent traceability drivers are adopted from Islam and Cullen (2021) and are discussed in the following subsections.

### *2.1 Legislation and certification*

Traceability has been embedded as an indispensable requirement in the food legislation of several countries, including the EU food law 178/2002; the US Bio-terrorism Act and the US Food Safety Modernization Act (FSMA); and Japan's Food Traceability Act (Bechini et al., 2008; Charlebois et al., 2014; Qian et al., 2020). Many of these food laws require the mandatory recording of the identification of immediate upstream suppliers and downstream customers, in what is known as the 'one up-one down' or 'one step forward-one step back' traceability approach.

A key driver motivating many food companies to execute traceability is to meet the necessary legal requirements so that they can stay functional in markets. Some countries agree to abide by traceability regulations to gain access to export markets and introduce their own guidelines. The traceability regulations developed by various parties (e.g. governments, regulators and market actors) are often heavily influenced by the standards developed by the International Organisation for Standardisation (ISO), such as ISO 22005 (ISO, 2005) which defines the functional requirements for the implementation of practical FTSs (Islam and Cullen, 2021). The market driver for independently verifiable FTSs has led to the development of third-party private certification schemes (Norton et al., 2014).

### *2.2 Safety and quality*

Demonstrating compliance with food safety and quality standards emerges as a strong motivating factor for implementing FTSs, driven by persistent shocks and costly disruptions to FSCs, including the European horse meat (scandal) substitution for beef, and more recently, ethylene oxide recalls across Europe and the COVID-19 pandemic. These incidents impact the perceived trustworthiness of FSCs and lead to calls for increased transparency surrounding the safety and quality of food products (Aung and Chang, 2014). The interaction between traceability and transparency in FSCs is gaining wider interest (Baralla et al., 2021; Feng et al., 2020), especially where public health has to be ensured within complex FSCs (Demestichas et al., 2020).

## *2.3 Customer satisfaction*

Food traceability systems both assure and create consumer confidence in the safety and quality of food (Qian et al., 2020; Zhang et al., 2020), especially in a market where consumers have a high level of food safety awareness (Rodriguez-Salvador and Dopico, 2020). The traceability protocols and the designs of FTSs to reduce food safety risks are framed by these consumer preferences (Food Standards Agency, 2002; Garaus and Treiblmaier, 2021). Indeed, Lam et al. (2020) argue that consumer trust in food safety is based on FTSs and information transparency.

## *2.4 Sustainability*

Traceability has become a strong basis for sustainability-related credence claims that are difficult for consumers to ascertain, such as whether a food has been produced through sustainable production practices (Golan et al., 2004). Companies increasingly make use of traceability to validate sustainability claims and gain a competitive advantage from responsible environmental and social practices (Norton et al., 2014; Gallo et al., 2021; Islam and Cullen, 2021; Islam et al., 2021b). For example, traceability helps companies to authenticate that the good animal welfare practices they claim to adhere to are in place, thereby leading to increased organisational reputation (Golan et al., 2004). With regard to fish supply chains in particular, illegal, unregulated and unreported (IUU) fishing can take place without concern for the environment or strict regulations and fishing quotas. Organisations can make use of FTSs to prevent IUU fish entering fish supply chains (Borit and Olsen, 2012). The Marine Stewardship Council (MSC) Standard for seafood sustainability requires batch-level traceability, confirming that there is no presence of IUU fish, from raw produce through to final consumer (Norton et al., 2014; Lin et al., 2020; Islam and Cullen, 2021). Thus, effective FTSs can underpin sustainability claims with regard to a given product.

## *2.5 Value and efficiency*

Inventory management, effective product recall and product distribution are important motivators for companies implementing FTSs (Islam and Cullen,

2021). Retailers make extensive use of traceability for efficient operational planning (Bourlakis and Bourlakis, 2006; Mishra et al., 2020); resource efficiency (Lin et al., 2020); optimised shelf-life based distribution planning (Jedermann et al., 2014); prevention of animal theft (Smith et al., 2005); and minimising product recall cost (Opara, 2003).

### 2.6 Summary

Together, the five drivers identified above - legislation and certification; safety and quality; customer satisfaction; sustainability; value and efficiency - underpin the case for improving traceability in FSCs and adopting new ICT technologies to deliver more efficient FTSs.

## 3 Food traceability in theory

Numerous definitions of food traceability exist, provided by various organisations and academics and detailed in multiple pieces of legislation (Islam and Cullen, 2021). The EU General Food Law (2002) defines food traceability as 'the ability to trace and follow a food, feed, food producing animal or substance intended to be, or expected to be incorporated into a food or feed, through all stages of production, processing and distribution'. According to the ISO 9000 (2000), 'traceability is the ability to trace the history, application or location of that which is under consideration'. Olsen and Borit (2013) present traceability as 'the ability to access any or all information relating to that which is under consideration, throughout its entire life cycle, by means of recorded identifications'. Islam and Cullen (2021) most recently define food traceability as 'an ability to access specific information about a food product that has been captured and integrated with the product's recorded identification throughout the supply chain'.

Moe (1998) provides further detail with the description of two main types of traceability in her definition: chain traceability (or external traceability), which is the 'ability to track a product batch and its history through the whole, or part, of a production chain from harvest through transport, storage, processing, distribution and sales', whereas internal traceability is the 'ability to trace ... in one of the steps in the chain'. This means that traceability can operate at two levels simultaneously, the supply chain level and the organisational level; and that for chain traceability to work effectively, all discrete internal traceability activities must be operationalised appropriately and synchronised with each other.

Internal traceability is enabled by the recording of product descriptors within the boundaries of a single food business operator (FBO), while external traceability requires transmission of that information to other links in the chain.

Internal traceability is executed by the grouping of raw materials and products into batches or lots and assigning them discrete identifiers (Olsen and Borit, 2018; Islam and Cullen, 2021). The batches or lots, alternatively known as traceable resource units (TRU) (Moe, 1998), undergo several transformations that may encompass aggregation (e.g. mixing and combining) or disaggregation (e.g. splitting and portioning). For example, harvested fruit from multiple growers is pooled as one TRU, or conversely, a livestock carcass being disaggregated to multiple cuts of meat, being distinct TRUs. As a TRU moves internally within an FBO and/or through the FSC, the associated information needs to be communicated in tandem. In essence, information flows through the FTS from material origin to final product, enabling traceability and visibility in supply chains. Thus, traceability, as described here, associates a product and its component parts; the organisations through which the product and its component parts have been transferred; the modifications/transformations that have occurred at each point; and the characteristics of the data itself that in turn give information about the product and any allied services.

## 4 Food traceability system

The implementation of traceability in FSCs, in practice, requires discrete FTSs. An FTS is the totality of data and operations that is capable of maintaining the desired information about the food product and its ingredients through all or part of the supply chain (World Economic Forum, 2019). Although paper-based manual recording is able to execute traceability in a one up-one down manner, it does not provide adequate transparency across the FSC. Transparency is defined as 'the practice of being open and honest with customers in terms of food sourcing, ingredient lists, nutrition facts, allergen concerns, and ecological impact' (FSA, 2017). Lack of transparency results in a greater risk of food adulteration, counterfeiting and mislabelling (Manning, 2019a). Therefore, computerised traceability systems including pedigree, centralised and decentralised data management approaches have emerged, which possess the potential to strengthen traceability leading to better transparency across FSCs. The evolution of traceability systems, from manual to computerised systems and across centralised and distributed systems, is discussed in Section 4.1.

### *4.1 Manual and computerised traceability systems*

Traceability systems have evolved from simple manual paper-based recordings to the latest Industry 4.0 technology-based information systems (Islam et al., forthcoming). Paper-based databases (Zhang et al., 2010) represent the simplest technique for data recording using paper trails, including the use of product registration books, faxes, paper forms and postal letters. Paper-based manual systems are adequate for implementing traceability systems based on the one up-one down approach to record the identification of immediate supplier and customer, and pedigree-based information systems, such as the audit trail recording of a product's transformation and ownership as it passes along the supply chain (Alfian et al., 2017; Olsen, 2017). Although paper-based traceability systems involving manual transcription require less expertise and less infrastructural cost to implement, they are time consuming, less reliable, less suitable for further analysis and highly prone to human error or data loss (Frederiksen and Bremner, 2001; Islam et al., 2021b).

Due to the weaknesses identified in paper-based traceability systems, computerised traceability systems offering improved accuracy and reliability have been extensively adopted across various FSCs (Regattieri et al., 2007; Zhang et al., 2010; Feng et al., 2013). Computerised databases can be updated by a manual transcription or an automated transcription through various technologies, for example, barcodes, RFID and wireless sensor networks (WSN). Computerised traceability systems for chain traceability across FSCs can be implemented through database software with electronic data interchange (EDI) capacity, centralised cloud-based information systems or decentralised blockchain technology. Although they require increased infrastructural costs, training and expertise, computerised traceability systems provide improved information storage and retrieval, reliability, transparency, data standardisation, security and ease of subsequent analysis and interrogation.

## 4.2 Central and distributed traceability system

Traceability systems can be categorised by the degree of independence of the databases used throughout the supply chain, that is, whether the traceability system is based on centralised or distributed information systems (Hugoson, 2007; Tian, 2017). An important distinction between centralised and decentralised information systems is the data location used in individual FBOs for recording and transmitting traceability data, which can be described as totally distributed, compressed and distributed, or centralised (Gandino et al., 2009). In the *totally distributed* approach, data are stored directly with details such as written words or RFID labels directly on the product or packaging, for example, the use of ear tags attached to the animals containing detailed animal history when the animal is moved from one FSC link to the next (Feng et al., 2013). In the *compressed and distributed* approach, data are stored in a compressed form (e.g. short code) which is linked to a reference database; for example, a barcode, containing product identification, links a product to a reference database which details product operation information (McCathie and Michael, 2005). However, all traceability data can also be stored in a centralised database and its link can be attached to the given item, say in the form of a QR code. Such an information system is commonly used in retail stores for product tagging, where the tag code links back to a centralised database containing detailed product information (Tarjan et al., 2014).

Another distinction between centralisation and decentralisation is observed in the choice to store data in a centralised or decentralised information system that influences the way traceability data are integrated or shared across multiple FBOs. In the centralised architecture, a central database owned by an external third party is responsible for all traceability data (Bechini et al., 2005). This architecture requires the responsible FBO to provide to a third party the information related to that activity (Huang et al., 2019). This information must allow the central database to, at a minimum, record the lot transformation, that is, the association between the input lot(s) and the output lot(s) (Bechini et al., 2005). This approach is, however, susceptible to single-node (central database) attack and data tampering, with the loss of data confidentiality leading to decreased data security (Huang et al., 2019).

In the decentralised approach, the FBO responsible for an activity records the transformation relationships between the associated input and output lots. The FBO then communicates the global identifier of their produced lot through a preferred communication channel (e.g., EDI, product tag or accompanying documents) to the distributor, which then enables the association of the input with the global identifier of the subsequent output (Bechini et al., 2005). This approach may suffer from lack of standardisation, information tampering and non-transparency (Tian, 2017).

Therefore, increased attention has recently been focused on an alternative decentralised data recording approach known as blockchain: a chronological chain of blocks with each block added to the chain after storing information regarding a distinct network activity (e.g., transaction, supply chain operation) with an identifiable timestamp (Tian, 2017). All users, considered as nodes in the network, are, with suitable permissions, able to see the recorded data in the chain of blocks at any time, but no one can change it, due to cryptographic protection (Salah et al., 2019). Being an immutable history of network activities, blockchain technology provides means to share traceability data among all nodes (i.e., supply chain actors) in the system. Using a blockchain-based information system removes the risk of using a centralised third-party traceability database, while providing transparency, authenticity, security and tamper-proofing of the shared traceability information for all nodes in the system. More detail about blockchain technology is provided in Section 5.7.

## 5 Traceability technologies

The so-called fourth industrial revolution (or Industry 4.0) is characterised by the ongoing automation of traditional manufacturing and industrial practices using modern smart technologies that enable large-scale machine-to-machine communication (M2M) and the Internet of Things (IoT) providing improved communication, self-monitoring and diagnosis of issues without the need for human intervention. The emergence of the Industry 4.0 revolution can also be seen in food traceability systems where automatic identification (auto-ID) technologies, sensors and communication technologies are integrated for managing food product information throughout FSCs (Chen, 2017; Islam et al., forthcoming). The emergence of these technologies in food traceability systems is shown in Table 1 and discussed in the following section.

### 5.1 Barcode

The barcode emerged in the early 1970s and is now considered to be the first auto-ID technology. It has been widely used in many different supply chain management practices (McCathie and Michael, 2005; Tu et al., 2020). A barcode is operationalised as a symbology that encodes information in the form of either one-dimensional (1D) parallel lines (i.e. bars and spaces) or two-dimensional (2D) matrices of dots (Palmer, 1989; Tu et al., 2020). Barcodes are printed on various types of materials which can be read by special dedicated optical barcode scanners, handheld personal digital assistants (PDAs) or smartphones using a special application (McCathie and Michael, 2005; Von Reischach et al., 2010; Tu et al., 2020). When the right infrastructure, software and hardware are in place, the automation provided by a barcode system greatly simplifies information collection, processing and tracking for the users (McCathie and Michael, 2005).

Over the years, many different kinds of 1D and 2D barcode symbologies have been developed, each with its own intended industrial purpose and characteristics, for example, Code 39, Code 128, Data Matrix and QR Code (Musa et al., 2014; Tu et al., 2020). Barcode scanning is considered a reliable form of production identification, owing to its higher accuracy rates and high speeds in comparison to manual data collection (McCathie and Michael, 2005). However, accuracy attained through barcode technology can be compromised as the technology requires humans to perform scans, leading to potential mistakes (McCathie and Michael, 2005). Barcode readers require direct 'line-of-sight' for barcode scanning and close-range reading of items one at a time, often making the data acquisition difficult and time consuming. This 'line-of-sight' property leads to more noticeable performance downturn for scanners in harsh environments, such as fog, dirt and rain (McCathie and Michael, 2005; Islam et al., 2021b). Barcodes are also more vulnerable to security loss, with the cloning or modification of 1D barcode tag information being much easier than, for example, RFID, whose information is not visible to the outside.

### 5.2 Radio frequency identification

Radio frequency identification is another auto-ID technology that uses radio frequency signals to identify products. It consists of three necessary hardware components: an electronic tag or transponder, a reader or interrogator and a supervising computer (Tan and Koo, 2014). The primary benefits of RFID over barcode technology are a large tag code offering item-level unique identification; rapid data acquisition simultaneously from multiple tags, which enables automatic TRU reading in the production line; non-line-of-sight and no-contact scanning, which allows tags to still be readable even when encapsulated; and security features, such as deep encryption and password protection, to mitigate against security breaches or counterfeiting (McCathie and Michael, 2005; Islam et al., 2021b).

The RFID technology is employed in a number of different systems, which vary by the reader transmission frequency, reading range, reading speed, operating environment and cost (RFID4U). The four basic frequency ranges are low frequency (LF, 125/134 KHz), high frequency (HF, 13.56 MHz), ultra-high frequency (UHF, 860–960 MHz) and microwave frequency (2.45 GHz). The RFIDs can also be categorised into passive RFID, semi-passive RFID and active RFID (Tu et al., 2020). Passive RFID tags do not possess their own power source and therefore can only send information when energised by encountering an RFID reader. This results in tags which are smaller, lightweight and less expensive; however, the lack of power supply limits the integration of some wireless sensors which require reliable continuous power to operate. Semi-passive RFID and active RFID tags, with on-tag batteries, provide longer reading ranges, higher data transmission rates and compatibility with a wide range of sensors for recording a product's physical and environmental data. The on-board power supply and additional functionalities increase the weight, size and cost of these RFID tags, in comparison to passive RFID transponders. The RFID tags are also classified by their tag memory programmability, which include Read-Only; Write Once, Read Many (WORM); and Read-Write (Islam et al., 2021b).

### 5.3 Near-field communication

Near-field communication (NFC) technology also uses radio frequency for data transmission and operates on the same frequency as HF RFID (13.56 MHz), supporting high data transfer rates of 106, 212, 424 or 848 Kbit/s (Mainetti et al., 2013). The NFC technology is capable of two-way communications, including data transfer between two NFC mobiles, and one-way communication, with NFC devices and NFC tags, in an arrangement similar to passive RFID. The NFC tags do not require a dedicated reader for data transfer; they can be

**Table 1** Traceability technologies with advantages and disadvantages

| Technologies | Technology type | Advantage | Disadvantage |
|---|---|---|---|
| Barcode | Identification technology | • Cheaper compared to RFID and manual paper document<br>• Easy to use<br>• More reliability, higher speed and accuracy than manual methods | • Line-of-sight data reading<br>• Risk of optical damage<br>• Less accuracy compared to RFID<br>• Time consuming for one-by-one tag data reading<br>• Shorter reading range than RFID<br>• More vulnerable to security loss in comparison to RFID |
| Radio frequency Identification (RFID) | Identification technology | • Line of sight is not required<br>• Longer reading range, high-speed reading and high accuracy<br>• Store larger amount of data<br>• Tag durability<br>• Multiple tag reading at a time increases automation<br>• Higher data security than barcode | • Not profitable for low-value items<br>• Cost of infrastructure is high<br>• High expertise or training required<br>• Tag collision may occur |
| Near-Field Communication (NFC) | Identification technology | • Line-of-sight data reading is not required<br>• Carries more data than barcode<br>• Wireless data transfer<br>• No need of dedicated reader as mobile phones can be used<br>• Offers read-write programmability for tag data | • Not profitable for low-value items<br>• High expertise or training required<br>• Slower reading rate than RFID as multiple tag reading is not possible<br>• Technology has not matured yet |

| | | | |
|---|---|---|---|
| Wireless Sensor Network (WSN) | Physical and environmental property recording technology | • Wireless product physical and environmental data collection<br>• Higher reading range than RFID<br>• Cooperation among multiple data collection nodes | • Cannot provide product identification<br>• Expensive for low-value items<br>• High expertise or training required |
| EPCglobal | Identification and communication standards | • Standards for product identification code and database<br>• Interoperability in external traceability information communication<br>• Easily scalable | • No technology standards are defined<br>• Difficult to adopt for manual traceability system<br>• No standards for sensor data |
| Cloud computing | Communication technology | • High scalability<br>• Agility and high availability<br>• Multi-sharing of databases<br>• Services in the pay-per-use mode<br>• Automated backup of traceability database | • Cannot provide data authentication<br>• Vulnerable to single-node attack<br>• May suffer from information tampering and label cloning<br>• Not possible to delegate accountability for data<br>• May encounter loss of confidentiality and malicious attack |
| Blockchain | Communication technology | • Provides transparency, trust and authenticity<br>• Does not require centralised third party<br>• Eliminates the requirements for large data sharing on product tag | • Slow information recording<br>• Susceptible to system hacking<br>• Expensive for low-value items and small holders<br>• Difficulty with technology scalability<br>• High expertise or training required |

read by any smartphone that has NFC capability. NFC-compatible phones are designed to send out short magnetic field pulses at the designated radio frequency, inducing an electric current in the tag's circuit, and relaying the information to the phone (Trafton, 2014). Like RFID, multiple NFC tags can be read at a time and they are capable of read-write programmability (Islam et al., 2021b).

## 5.4 Wireless sensor network

A WSN is a network of spatially dispersed and dedicated sensors that collects sensing data of various physical or environmental conditions, and communicates them with a base station or central node, from where the data are then transmitted to a database server using an IoT gateway and mobile networks; for example, General Packet Radio Service (GPRS) (Costa et al., 2013; Aung and Chang, 2014). The WSN is built having between a few to several hundred 'nodes', where each node, consisting of a radio transceiver with an antenna connection, a microcontroller, memory capacity and energy source, connects to one or several sensors through an electronic circuit.

Although WSN nodes have limited processing capability and memory (Hammervoll et al., 2012), their ability to communicate with each other in various network topologies offsets the impact of missing communication from any single node to the base station, and hence reduces error rate (Costa et al., 2013). This is the main advantage of WSN over RFID systems, as RFID devices do not possess cooperative capabilities. The WSNs have much longer reading ranges than RFID, due to their ability to relay information from node to node, but are not suitable for identification purposes owing to their inability to transmit product identification (Ruiz-Garcia et al., 2009; Aung and Chang, 2014). A further advantage for the WSN system is the feasibility of installation in places where cabling is impossible, such as within the cargo for monitoring physical and environmental conditions of perishable food (Ruiz-Garcia et al., 2009).

At a reasonable cost and power consumption, WSN can provide real-time information on various parameters, such as the temperature, humidity and vibration experienced by perishable food in transit (Hammervoll et al., 2012). Power consumption can be reduced by placing the sensors in sleep mode after each successful data emission, and by activating the sensor only when required, at regular intervals (Aung and Chang, 2014). Low temperature is known to have a negative effect on the battery life of sensors (Ruiz-Garcia et al., 2009). The ZigBee communication protocol has been proposed to prolong the battery life of sensors. ZigBee is built on the IEEE 802.15.4 standard and supports various network topologies for WSNs with low energy consumption and low data transmission rates (Óskarsdóttir and Oddsson, 2019).

## 5.5 EPCglobal standards

EPCglobal is a worldwide association of supply chain partners providing open standards for Electronic Product Code (EPC) and Electronic Product Code Information System (EPCIS). EPCglobal enables effective communication through recording and exchange of product data as a product moves through supply chain actors and across country borders (Asher et al., 2007). The EPC is a unique, serialised identification code format for any kind of product that can be encoded from familiar identifiers, such as Global Trade Item Number (GTIN) and Serialised Shipping Container Code (SSCC). The EPCIS is a universal data model for representing the location and state of the product as it moves across organisational boundaries, while providing a means for communication between these organisations regardless of the technology suppliers.

The main components of an EPCIS data model include EPC, Event Time, Business Step, Disposition, Read Point, Business Location and Business Transaction. However, it is often extended in practice to also include Expiration Date, Batch Number and Temperature. The EPCIS standard is data carrier neutral and can be used to describe data captured from various technologies including RFID tags, barcodes and other data carriers (Asher et al., 2007). The EPCIS databases of different supply chain participants, being connected via local networks or the internet, provide a lineage of product movement which is accessible by authorised parties. The accessibility is enabled by deploying a discovery mechanism (e.g. object naming service or ONS) that takes an EPC as an input and replies with the address (in the form of a Uniform Resource Locator, or URL) of an EPCIS repository implemented by the supply chain actor corresponding to the EPC in question (Musa et al., 2014).

The EPCIS framework for an industry is extensible, in that it allows new industry requirements to co-exist alongside the core components while still providing interoperability (Asher et al., 2007). There are no technology implementation standards for EPCIS and there are many solution providers who provide EPCIS-compliant technology solutions, using firewalls, encoding and other safe measures to guarantee the security of the transmitted information (Musa et al., 2014).

## 5.6 Cloud computing

Cloud computing is the provision of dynamically scalable and often virtualised resources, for example, data storage or software application as a service (SAAS), over the internet (Patil et al., 2012). In practice, cloud computing often refers to internet-based development and the use of computing technology. Clouds may be limited to a single organisation, such as enterprise clouds managed by the enterprise IT department, or can be available to multiple organisations, for example public clouds managed by a third party (Wang et al., 2012). The cloud computing architecture comprises three layers: resource, platform and application. The resource layer consists of various hardware used for identification and communication, including PDAs, computers and middleware, which are owned by supply chain operators. Middleware can be described as a software that is the glue or the hidden transition layer bridging the gaps between applications, tools and databases in order to provide unified services to users (Microsoft Azure, n.d.). The platform layer includes components such as web servers, application servers and enterprise service buses and serves users by providing functionality such as information access. The application layer is the main layer consisting of the applications to serve the users, such as various internet-based software programs. Cloud computing provides several advantages for FTSs, including increased scalability, agility, availability, multi-sharing, services in a pay-per-use mode, automated backup and support for service-oriented applications (Wang et al., 2012).

At present, the mainstream traceability systems for food products are often maintained and governed by the government or a leading enterprise by using a centralised cloud-based information system, which participants with accessibility permission can use to input their respective traceability information or to simply access to visualise the data. Such an approach facilitates a common standard database for users, reducing the need for data standardisation efforts by individual companies. However, cloud-based systems may suffer from information tampering, label cloning, product counterfeiting, illegal product trading, accountability difficulty, loss of confidentiality and malicious attack of the central database (Huang et al., 2019; Manning, 2019b).

## 5.7 Blockchain

Unlike cloud databases managed by central authorities, blockchain technology provides a distributed database that is participated in, and maintained by, all nodes in a peer-to-peer (P2P) network, or all actors in a FSC network (Li et al., 2020). The blockchain is an immutable, distributed ledger. In practice, this is an ever-growing list of chronologically linked records, called blocks, which store transaction and asset data across many computers throughout the supply chain network. Each block also contains a cryptographic hash of the previous block and a timestamp, so that any involved block in a given connection cannot be altered retroactively, without the associated alteration of all subsequent blocks (Wang et al., 2018). Due to its tamper-proof, secure and traceable nature, blockchains can be deployed effectively in implementing FTSs (Salah et al., 2019). Blockchain is often implemented by deploying a smart contract: a computer program which defines the protocols of a business contract that governs and controls all interactions and transactions among all the participants involved within a P2P network or supply chain, without the need for a centralised third party (Salah et al., 2019). Blockchain-based FTSs mainly record and merge the product traceability data and transaction information between FSC parties. Each company in the network also maintains their own individual internal traceability database, often using decentralised file systems, which allows linking of internal traceability data with external traceability data in the block and the verification of traceability data integrity (Salah et al., 2019; Li et al., 2020).

Product traceability data are typically stored in decentralised file systems and uploaded to the smart contract platform and then broadcast over the whole blockchain network to ensure authenticity (Huang et al., 2019). Data are verified by the consensus mechanism by other nodes in the blockchain network before being stored. Trade requests are first verified before the product ownership is transferred, and any transaction information is packed into a block. Each block is hashed and linked to the next block, making it a secure chain of immutable and tamper-proof records of all transactions, spread across all nodes in the network. In this distributed information system, smart contracts ensure the standardisation of the recorded traceability data (Li et al., 2020). Ethereum, a programmable smart contract platform, is popularly used for implementing blockchain-based information systems among supply chains.

The advantages of blockchain-based traceability systems include decentralisation, tamper-proof protection, data verification, privacy, distributed data storage, transparency and increased trust (Abeyratne and Monfared, 2016; Tian, 2017; Behnke and Janssen, 2020). However, the technology still needs some improvement as it suffers from various disadvantages: the blockchain network can become bulky with time if an unrestricted number of users join, slowing down the addition of new data (Behnke and Janssen, 2020); some blockchain designs do not have any functionality to protect sensitive information from certain users (Feng et al., 2019); blockchain software written with poorly developed codes is vulnerable to hacking (DeVries, 2016); blockchain traceability systems can incur high implementation and electricity consumption costs (Li et al., 2019); there is a high requirement for training (Behnke and Janssen, 2020); and the technology is not profitably scalable across small supply chains (World Economic Forum, 2019) (Table 1).

## 6 Barriers in implementation of effective food traceability systems

Despite the enormous potential of emerging technologies to improve traceability, various barriers exist that hinder their implementation in FTSs. Some of these barriers are discussed in the following subsections.

### *6.1 Lack of awareness*

There is a lack of understanding of the theoretical traceability concepts and of how traceability differs from other similar concepts such as chain of custody (Olsen, 2017). For example, a UK survey of small- and medium-sized food and drink industries shows that only 25% of participant companies align the term 'traceability' with information and product logistics, leading to reduced effectiveness of their traceability systems and implementation (Mattevi and Jones, 2016). In addition, there appears to be widespread resistance to the implementation and use of new technologies, with many FBOs still relying on manual traceability systems. Many FBOs are not aware of all the potential benefits of implementing improved traceability and consider only the additional upfront capital investment and legal requirements. Instead, they persist with paper-based one up-one down traceability systems, which are limited to only the identification of suppliers and customers (Asioli et al., 2014).

### *6.2 Resource deficiencies*

A barrier to improved practical food traceability results from resource deficiencies and a lack of funding, as traceability implementation is expensive and complicated (Bosona and Gebresenbet, 2013). During traceability implementation, the allocation of costs and benefits among FSC partners requires considerable effort, and results in initial resistance against implementation. The introduction of new FTSs also requires increased administrative work, training and capacity building, especially when companies are implementing traceability systems for the first time (Bosona and Gebresenbet, 2013).

### *6.3 Inefficient data management*

Inefficient information flow in FTSs has been revealed, in practice, by recurrent food crises (Badia-Melis et al., 2015). Inefficient information flow is associated with poor traceability data management in the internal FTSs of individual FBOs (Duan et al., 2017). Information loss for an FBO occurs when the product and its operational or quality information are not systematically recorded or linked at the key information collection points (Karlsen et al., 2011; Zhang and Bhatt, 2014). This can happen as a result of failure in TRU identification and transformation recording, incompetent recording techniques, absence of industry-specific standard data lists, or simply human error (Bertolini et al., 2006; Karlsen et al., 2011; Zhang and Bhatt, 2014; Karlsen and Olsen, 2016).

### *6.4 Lack of standards*

Various FSC operators generate an increasing amount of traceability data using various technologies, essential for effective traceability. For example, a farm may use simple paper trails for data recording, while processors are equipped with enterprise resource planning (ERP) software. However, the prevalence of heterogeneous data sources means that a traceability system may face significant interoperability challenges and data integration issues (Bougdira et al., 2019). This problem is amplified by the lack of a widely adopted standard for how FBOs record and transmit data for traceability, which, in turn, hinders the adoption of novel communication technologies (e.g., blockchain) (World Economic Forum, 2019).

### 6.5 Technology cost and scaling issues

Cost is often considered to be the main constraint preventing the more widespread use of ICT and Industry 4.0 technologies (e.g. RFID, WSN, blockchain), especially in the case of low-value items procured from smallholders (Islam et al., forthcoming). Some technologies (e.g. blockchain) do not appear profitable for small-scale FSCs due to their high set-up cost (World Economic Forum, 2019). However, the upfront cost of implementing improved traceability systems using novel technologies is frequently considered in isolation as an additional cost, without weighing it against the potential cost reduction through ensuring the safety, quality and authenticity of food products. This is because the benefits of improved traceability and the reduction in risk to the food supply are challenging to price. There is an urgent need for scientists and practitioners to co-develop new tools to assess the risks and potential costs of not implementing more advanced traceability systems.

## 7 Improvement pathways

To adopt new technologies and achieve desired outcomes, new pathways need to be created to drive improved traceability. Some improvement pathways and policy interventions are outlined here:

- Multi-stakeholder collaborations combined with financing and policy incentives need to be developed that can support FBOs in adopting emerging technologies that would otherwise necessitate high capital investments and operational costs beyond the reach of such organisations (World Economic Forum, 2019).
- Ongoing technological development needs to be supported through policy intervention to drive down the costs of adopting more advanced traceability solutions. An example can be the 'Contract for Difference (CfD)' of the UK government to reduce prices of renewable electricity technologies (Evans, 2017).

- Technology and data format standards are available for national animal traceability in many countries (Charlebois et al., 2014). Similar standards from GS1 can be embedded in the legal requirements for traceability of foods which are more vulnerable to fraud (e.g., olive oil and seafood) (Mermelstein, 2015).
- Sector-specific clear, consistent and globally harmonised standards should continue to be developed, such as ISO 12875:2011 and 12877:2011, for captured and farmed finfish (Olsen, 2017).
- New tools and frameworks for assessing the risks and potential costs of food quality loss and safety breaches against the investment costs in advanced traceability systems are required to be developed, to encourage more advanced FTSs.

## 8 Where to look for further information

The following articles provide a good overview of the subject discussed in this chapter:

- Islam, S. and Cullen, J. M. (2021). Food traceability: a generic theoretical framework, Food Control, 123, 107848.
- Badia-Melis, R., Mishra, P. and Ruiz-García, L. (2015). Food traceability: new trends and recent advances: a review, Food Control, 57, 393–401.
- Bosona, T. and Gebresenbet, G. (2013). Food traceability as an integral part of logistics management in food and agricultural supply chain, Food Control, 33(1), 32–48.
- Lees, M. (ed.) (2003). Food Authenticity and Traceability. Elsevier.
- World Economic Forum (2019). Innovation with a purpose: improving traceability in food value chains through technology innovations.

## 9 References


Abeyratne, S. A. and Monfared, R. P. (2016). Blockchain ready manufacturing supply chain using distributed ledger, *International Journal of Research in Engineering and Technology* 5(9), 1–10.

Alfian, G., Rhee, J., Ahn, H., Lee, J., Farooq, U., Ijaz, M. F. and Syaekhoni, M. A. (2017). Integration of RFID, wireless sensor networks, and data mining in an e-pedigree food traceability system, *Journal of Food Engineering* 212, 65–75. doi: 10.1016/j.jfoodeng.2017.05.008.

Alfian, G., Syafrudin, M., Farooq, U., Ma'arif, M. R., Syaekhoni, M. A., Fitriyani, N. L., Lee, J. and Rhee, J. (2020). Improving efficiency of RFID-based traceability system for perishable food by utilizing IoT sensors and machine learning model, *Food Control* 110, 107016. doi: 10.1016/j.foodcont.2019.107016.

Asher, C., Morgan, G., Swan, R. and Traub, K. (2007). *EPCIS (Electronic Product Code Information Service) Frequently Asked Questions, EPCglobal*. https://www.gs1.org/docs/epc/epcis_1_0-faq-20070427.pdf.

Asioli, D., Boecker, A. and Canavari, M. (2014). On the linkages between traceability levels and expected and actual traceability costs and benefits in the Italian fishery supply chain, *Food Control*. Elsevier 46, 10–17.

Aung, M. M. and Chang, Y. S. (2014). Traceability in a food supply chain: safety and quality perspectives, *Food Control*. Elsevier Ltd 39, 172–184. doi: 10.1016/j.foodcont.2013.11.007.

Badia-Melis, R., Mishra, P. and Ruiz-García, L. (2015). Food traceability: new trends and recent advances: a review, *Food Control* 57, 393–401. doi: 10.1016/j.foodcont.2015.05.005.

Baralla, G., Pinna, A., Tonelli, R., Marchesi, M. and Ibba, S. (2021). Ensuring transparency and traceability of food local products: a blockchain application to a Smart Tourism Region, *Concurrency and Computation: Practice and Experience* 33(1), e5857.

Bechini, A., et al. (2005). A general framework for food traceability. In: 2005 Symposium on Applications and the Internet Workshops (SAINT 2005 Workshops), IEEE Publications, pp. 366–369.

Bechini, A., Cimino, M. G., Marcelloni, F. and Tomasi, A. (2008). Patterns and technologies for enabling supply chain traceability through collaborative e-business, *Information and Software Technology*, 50(4), 342–359.

Behnke, K. and Janssen, M. F. W. H. A. (2020). Boundary conditions for traceability in food supply chains using Blockchain technology, *International Journal of Information Management*. Elsevier 52, 101969.

Bertolini, M., Bevilacqua, M. and Massini, R. (2006). FMECA approach to product traceability in the food industry, *Food Control* 17(2), 137–145. doi: 10.1016/j.foodcont.2004.09.013.

Blank, C. (2018). FDA's shrimp import refusals rose in January, Seafoodsource. Available at: https://www.seafoodsource.com/news/food-safety-health/fdas-shrimp-import-refusals-rose-in-january (Accessed: 18 January 2021).

Borit, M. and Olsen, P. (2012). Evaluation framework for regulatory requirements related to da-ta recording and traceability designed to prevent illegal, unreported and unregulated fishing, *Marine Policy*, 36(1), 96–102. http://doi.org/10.1016/j.marpol.2011.03.012.

Bosona, T. and Gebresenbet, G. (2013). Food traceability as an integral part of logistics management in food and agricultural supply chain, *Food Control* 33(1), 32–48. doi: 10.1016/j.foodcont.2013.02.004.

Bougdira, A., Akharraz, I. and Ahaitouf, A. (2019). A traceability proposal for industry 4.0, *Journal of Ambient Intelligence and Humanized Computing* 11(8), 3355–3369.

Bourlakis, M. and Bourlakis, C. (2006). Integrating logistics and information technology strategies for sustainable competitive advantage, *Journal of Enterprise Information Management*. Emerald Group Publishing Limited 19(4), 389–402.

Charlebois, S., Sterling, B., Haratifar, S. and Naing, S. K. (2014). Comparison of global food traceability regulations and requirements, *Comprehensive Reviews in Food Science and Food Safety*. Wiley Online Library 13(5), 1104–1123.

Chen, R. Y. (2017). An intelligent value stream-based approach to collaboration of food traceability cyber physical system by fog computing, *Food Control*. Elsevier Ltd 71, 124–136. doi: 10.1016/j.foodcont.2016.06.042.

Costa, C., Antonucci, F., Pallottino, F., Aguzzi, J., Sarriá, D. and Menesatti, P. (2013). A review on agri-food supply chain traceability by means of RFID technology, *Food and Bioprocess Technology*. Springer 6(2), 353–366.

Demestichas, K., Peppes, N., Alexakis, T. and Adamopoulou, E. (2020). Blockchain in agriculture traceability systems: a review, *Applied Sciences* 10(12), 4113.

DeVries, P. D. (2016). An analysis of cryptocurrency, bitcoin, and the future, *International Journal of Business Management and Commerce* 1(2), 1–9.

Duan, Y., Miao, M., Wang, R., Fu, Z. and Xu, M. (2017). A framework for the successful implementation of food traceability systems in China, *Information Society* 33(4), 226–242. doi: 10.1080/01972243.2017.1318325.

El Maouchi, M., Ersoy, O. and Erkin, Z. (2018). TRADE: a transparent, decentralized traceability system for the supply chain. In: *Proceedings of the 1st ERCIM Blockchain Workshop*, European Society for Socially Embedded Technologies (EUSSET).

EU (2002). Regulation (EC) no 178/2002 of the European Parliament and of the Council of 28 January 2002 laying down the general principles and requirements of food law, establishing the European Food Safety Authority and laying down procedures in matters of food safety, *Official Journal of the European Communities* 31, 1–24.

Evans, S. (2017). Analysis: UK auction reveals offshore wind cheaper than new gas, CarbonBrief. Available at: https://www.carbonbrief.org/analysis-uk-auction-offshore-wind-cheaper-than-new-gas (Accessed 21 January 2021).

FAO (2019). The future of food safety, First FAO/WHO/AU International Food Safety Conference Addis. Available at: http://www.fao.org/3/ca4289en/CA4289EN.pdf (Accessed 12–13 February 2019).

Feng, H., Wang, X., Duan, Y., Zhang, J. and Zhang, X. (2020). Applying blockchain technology to improve agri-food traceability: a review of development methods, benefits and challenges, *Journal of Cleaner Production* 260, 121031.

Feng, J. et al. (2013). Development and evaluation on a RFID-based traceability system for cattle/beef quality safety in China, *Food Control*. Elsevier Ltd 31(2), 314–325. doi: 10.1016/j.foodcont.2012.10.016.

Feng, Q., He, D., Zeadally, S., Khan, M. K. and Kumar, N. (2019). A survey on privacy protection in blockchain system, *Journal of Network and Computer Applications* 126, 45–58.

Food Standards Agency (2002). *Traceability in the Food Chain. A Preliminary Study*, Food Chain Strategy Division, Food Standards Agency, London.

Food Standards Agency (2017). *Food Standards Agency Transparency: Understanding Public Views and Priorities Report*.

Frederiksen, M. T. and Bremner, A. (2001). Fresh fish distribution chains: an analysis of three Danish and three Australian chains, *Food Australia*. Australian Institute of Food Science and Technology Inc. 54(4), 117–123.

Gallo, A., Accorsi, R., Goh, A., Hsiao, H. and Manzini, R. (2021). A traceability-support system to control safety and sustainability indicators in food distribution, *Food Control* 124, 107866.

Gandino, F., Montrucchio, B., Rebaudengo, M. and Sanchez, E. R. (2009). On improving automation by integrating RFID in the traceability management of the agri-food sector, *IEEE Transactions on Industrial Electronics*. IEEE Publications 56(7), 2357–2365.

Garaus, M. and Treiblmaier, H. (2021). The influence of blockchain-based food traceability on retailer choice: the mediating role of trust, *Food Control* 129, 108082.

Golan, E., et al. (2004). Traceability in the U. S. food supply: economic thecry and industry studies, traceability in the U. S. Food supply: economic theory and industry studies. doi: 10.1001/jama.1971.03190220052018.

Hammervoll, T., Jensen, L. M., Hafliðason, T., Ólafsdóttir, G., Bogason, S. and Stefánsson, G. (2012). Criteria for temperature alerts in cod supply chains, *International Journal of Physical Distribution and Logistics Management*. Emerald Group Publishing Limited, 42, 355–371.

Hardt, M. J., Flett, K. and Howell, C. J. (2017). Current barriers to large-scale interoperability of traceability technology in the seafood sector, *Journal of Food Science*. Wiley Online Library 82(S1), A3–A12.

Huang, H., Zhou, X. and Liu, J. (2019). Food supply chain traceability scheme based on Blockchain and EPC technology. In: *Lecture Notes in Computer Science*, International Conference on Smart Blockchain. Springer, 11911, pp. 32–42.

Hugoson, M.-Å. (2007). Centralized versus decentralized information systems. In: *IFIP Advances in Information and Communication Technology*, IFIP Conference on History of Nordic Computing. Springer, pp. 106–115.

Islam, S. and Cullen, J. M. (2021). Food traceability: a generic theoretical framework, *Food Control*. Elsevier 123, 107848.

Islam, S., Cullen, J. M. and Manning, L. (2021a). Visualising food traceability systems: a novel system architecture for mapping material and information flow, *Trends in Food Science and Technology* 112, 708–719. doi: 10.1016/j.tifs.2021.04.020.

Islam, S., Manning, L. and Cullen, J. M. (2021b). A Hybrid Traceability Technology Selection approach for Sustainable Food Supply Chains, *Sustainability* 13(16), 9385.

Islam, S., Manning, L. and Cullen, J. M. (forthcoming). Selection criteria for planning cold food chain traceability technology enabling industry 4.0, Procedia computer science, 3rd International Conference on Industry 4.0 and Smart Manufacturing (ISM 2021), 17–19 November 2021, Austria.

ISO (2000). Quality Management Systems-Fundamentals and Vocabulary. ISO/TC 176/SC 1(9000), 2000.

ISO (2005). *Traceability in the Feed and Food Chain – General Principles and Basic Requirements for System Design and Implementation (2005).* ISO/TC 176/SC 1(22005).

Jedermann, R., Nicometo, M., Uysal, I. and Lang, W. (2014). Reducing food losses by intelligent food logistics, *Philosophical Transactions. Series A, Mathematical, Physical, and Engineering Sciences* 372(2017), 20130302. doi: 10.1098/rsta.2013.0302.

Karlsen, K. M., Donnelly, K. A.-M. and Olsen, P. (2011). Granularity and its importance for traceability in a farmed salmon supply chain, *Journal of Food Engineering*. Elsevier Ltd 102(1), 1–8. doi: 10.1016/j.jfoodeng.2010.06.022.

Karlsen, K. M. and Olsen, P. (2016). Problems and implementation hurdles in food traceability, advances in food traceability techniques and technologies: improving quality throughout the food chain, Elsevier Ltd. doi: 10.1016/B978-0-08-100310-7.00003-X.

Kendall, H., Clark, B., Rhymer, C., Kuznesof, S., Hajslova, J., Tomaniova, M., Brereton, P. and Frewer, L. (2019). A systematic review of consumer perceptions of food fraud and authenticity: a European perspective, *Trends in Food Science and Technology*. Elsevier 94(October), 79–90. doi: 10.1016/j.tifs.2019.10.005.

Lam, T. K., Heales, J., Hartley, N. and Hodkinson, C. (2020). Consumer trust in food safety requires information transparency, *Australasian Journal of Information Systems* 24, 1–28.

Li, J., Li, N., Peng, J., Cui, H. and Wu, Z. (2019). Energy consumption of cryptocurrency mining: a study of electricity consumption in mining cryptocurrencies, *Energy* 168, 160–168.

Li, Y., Chu, X., Tian, D., Feng, J. and Mu, W. (2020). A traceability architecture for the fresh food supply chain based on blockchain technology in China. In: *Communications in Computer and Information Science*, International Conference on Artificial Intelligence and Security. Springer, pp. 347-358.

Lin, K., Chavalarias, D., Panahi, M., Yeh, T., Takimoto, K. and Mizoguchi, M. (2020). Mobile-based traceability system for sustainable food supply networks, *Nature Food* 1(11), 673–679.

Mainetti, L., Patrono, L., Stefanizzi, M. L. and Vergallo, R. (2013). An innovative and low-cost gapless traceability system of fresh vegetable products using RF technologies and EPCglobal standard, *Computers and Electronics in Agriculture* 98, 146–157. doi: 10.1016/j.compag.2013.07.015.

Manning, L. (2019a). Corporate responsibility in a transitioning food environment: truth-seeking and truth-telling. In: Piatti C., Graeff-Hönninger S. and Khajehei F. (eds). *Food Tech Transitions*, Springer, Cham, pp. 149–169. https://doi.org/10.1007/978-3-030-21059-5_9.

Manning, L. (2019b). Food defence: refining the taxonomy of food defence threats, *Trends in Food Science and Technology* 85, 107–115.

Mattevi, M. and Jones, J. A. (2016). Food supply chain: are UK SMEs aware of concept, drivers, benefits and barriers, and frameworks of traceability?, *British Food Journal* 118(5), 1107–1128. doi: 10.1108/BFJ-07-2015-0261.

McCathie, L. and Michael, K. (2005). *Is It the End of Barcodes in Supply Chain Management?* In Proceedings of the Collaborative Electronic Commerce Technology and Research Conference LatAm, University of Talca, Talca, Chile, pp. 1–19.

McLinden, T., Sargeant, J. M., Thomas, M. K., Papadopoulos, A. and Fazil, A. (2014). Component costs of foodborne illness: a scoping review, *BMC Public Health*. Springer 14, 509.

Mermelstein, N. H. (2015). Fighting food fraud, Institute of Food Technologists, pp. 67–73. Available at: https://www.ift.org/~/media/food%20technology/pdf/2015/03/0315_col_foodsafetyquality.pdf (Accessed 20 January 2021).

Microsoft (n.d.). Azure, What Is Middleware? Available at: https://azure.microsoft.com/en-us/overview/what-is-middleware/ (Accessed 10 April 2021).

Mishra, N., Mistry, S., Choudhary, S., Kudu, S. and Mishra, R. (2020). Food traceability system using blockchain and QR code. In: *IC-BCT 2019*. Springer, Singapore, pp. 33–43.

Moe, T. (1998). Perspectives on traceability in food manufacture, *Trends in Food Science and Technology* 9(5), 211–214. doi: 10.1016/S0924-2244(98)00037-5.

Musa, A., Gunasekaran, A. and Yusuf, Y. (2014). Supply chain product visibility: methods, systems and impacts, *Expert Systems with Applications*. Elsevier 41(1), 176–194.

Naylor, N. R., et al. (2017). A national estimate of the health and cost burden of *Escherichia coli* bacteraemia in the hospital setting: the importance of antibiotic resistance, *bioRxiv*. Cold Spring Harbor Laboratory, 153775.

Norton, T., et al. (2014). A guide to traceability, United Nations Global Compact Office, p. 45. Available at: https://www.bsr.org/reports/BSR_UNGC_Guide_to_Traceability.pdf.

Olsen, P. (2017). Food traceability in theory and in practice. Available at: https://ninum.uit.no/handle/10037/15408.

Olsen, P. and Borit, M. (2013). How to define traceability, *Trends in Food Science and Technology*. Elsevier 29(2), 142–150.

Olsen, P. and Borit, M. (2018). The components of a food traceability system, *Trends in Food Science and Technology*. Elsevier 77, 143–149. doi: 10.1016/j.tifs.2018.05.004.

Opara, L. U. (2003). Traceability in agriculture and food supply chain: a review of basic concepts, technological implications, and future prospects. *Journal of Food Agriculture and Environment*, 1, 101–106. https://pdfs.semanticscholar.org/4f63/2b6ec4bd1ebdc53a0fcf295b1cd3800ed11e.pdf.

Óskarsdóttir, K. and Oddsson, G. V. (2019). Towards a decision support framework for technologies used in cold supply chain traceability, *Journal of Food Engineering*. Elsevier 240, 153–159.

Palmer, R. C. (1989). *The Bar Code Book: Reading, Printing, and Specification of Bar Code Symbols*. Helmers Publishing.

Patil, V. C., et al. (2012). Internet of Things (Iot) and cloud computing for agriculture: an overview. *Proceedings of the Agro-Informatics and Precision Agriculture*. AIPA, India, pp. 292–296.

Pearson, S., May, D., Leontidis, G., Swainson, M., Brewer, S., Bidaut, L., Frey, J. G., Parr, G., Maull, R. and Zisman, A. (2019). Are distributed ledger technologies the panacea for food traceability?, *Global Food Security*. Elsevier B.V. 20, 145–149. doi: 10.1016/j.gfs.2019.02.002.

Qian, J., Ruiz-Garcia, L., Fan, B., Villalba, J. I. R., McCarthy, U., Zhang, B., Yu, Q. and Wu, W. (2020). Food traceability system from governmental, corporate, and consumer perspectives in the European Union and China: a comparative review, *Trends in Food Science and Technology* 99, 402–412.

Regattieri, A., Gamberi, M. and Manzini, R. (2007). Traceability of food products: general framework and experimental evidence, *Journal of Food Engineering* 81(2), 347–356. doi: 10.1016/j.jfoodeng.2006.10.032.

RFID4U. How to select a correct tag – frequency. Available at: https://rfid4u.com/rfid-frequency/ (Accessed 29 December 2020).

Rodriguez-Salvador, B. and Dopico, D. C. (2020). Understanding the value of traceability of fishery products from a consumer perspective, *Food Control*. Elsevier 112, 107142.

Ruiz-Garcia, L., Lunadei, L., Barreiro, P. and Robla, J. I. (2009). A review of wireless sensor technologies and applications in agriculture and food industry: state of the art and current trends, *Sensors* 9(6), 4728–4750. doi: 10.3390/s90604728.

Salah, K., Nizamuddin, N., Jayaraman, R. and Omar, M. (2019). Blockchain-based soybean traceability in agricultural supply chain, *IEEE Access* 7, 73295–73305. doi: 10.1109/ACCESS.2019.2918000.

Smith, G. C., Tatum, J. D., Belk, K. E., Scanga, J. A., Grandin, T. and Sofos, J. N. (2005). Traceability from a US perspective, *Meat Science* 71(1), 174–193. doi: 10.1016/j.meatsci.2005.04.002.

Sterling, B., Gooch, M., Dent, B., Marenick, N., Miller, A. and Sylvia, G. (2015). Assessing the value and role of seafood traceability from an entire value-chain perspective, *Comprehensive Reviews in Food Science and Food Safety*. Wiley Online Library 14(3), 205–268.

Tan, J. and Koo, S. G. M. (2014). A survey of technologies in Internet of Things. In: 2014 IEEE International Conference on Distributed Computing in Sensor Systems. IEEE Publications, pp. 269–274.

Tarjan, L., Šenk, I., Tegeltija, S., Stankovski, S. and Ostojic, G. (2014). A readability analysis for QR code application in a traceability system, *Computers and Electronics in Agriculture*. Elsevier 109, 1–11.

Thakur, M. and Forås, E. (2015). EPCIS-based online temperature monitoring and traceability in a cold meat chain, *Computers and Electronics in Agriculture* 117, 22–30. doi: 10.1016/j.compag.2015.07.006.

Tian, F. (2017). A supply chain traceability system for food safety based on HACCP, Blockchain & Internet of Things. In: 2017 International conference on service systems and service management. IEEE Publications, pp. 1–6.

Trafton, A. (2014). Detecting gases wirelessly and cheaply. Available at: https://news.mit.edu/2014/wireless-chemical-sensor-for-smartphone-1208 (Accessed 31 January 2021). MIT Press.

Tu, Y. J., Zhou, W. and Piramuthu, S. (2020). Critical risk considerations in auto-ID security: barcode vs. RFID, *Decision Support Systems* 142(December). doi: 10.1016/j.dss.2020.113471.

Von Reischach, F., Karpischek, S., Michahelles, F. and Adelmann, R. (2010). Evaluation of 1D barcode scanning on mobile phones, In the proceedings *Internet of Things (IOT)*, 1–5, IEEE.

Wang, H., He, W. and Wang, F. K. (2012). Enterprise cloud service architectures, *Information Technology and Management* 13(4), 445–454. doi: 10.1007/s10799-012-0139-4.

Wang, J., Li, M., He, Y., Li, H., Xiao, K. and Wang, C. (2018). A Blockchain-based privacy-preserving incentive mechanism in crowdsensing applications, *IEEE Access*. IEEE Publications 6, 17545–17556.

Wilcock, A., Pun, M., Khanona, J. and Aung, M. (2004). Consumer attitudes, knowledge and behaviour: a review of food safety issues, *Trends in Food Science and Technology*. Elsevier 15(2), 56–66.

World Economic Forum (2019). *Innovation with a Purpose: Improving Traceability in Food Value Chains through Technology Innovations*. https://www3.weforum.org/docs/WEF_Traceability_in_food_value_chains_Digital.pdf.

World Health Organization (2015). *WHO estimates of the global burden of foodborne diseases: foodborne disease burden epidemiology reference group 2007-2015 (No. 9789241565165).*

Zhang, A., Mankad, A. and Ariyawardana, A. (2020). Establishing confidence in food safety: is traceability a solution in consumers' eyes?, *Journal of Consumer Protection and Food Safety* 15(2), 99–107.

Zhang, J. and Bhatt, T. (2014). A guidance document on the best practices in food traceability, *Comprehensive Reviews in Food Science and Food Safety* 13(5), 1074–1103. doi: 10.1111/1541-4337.12103.

Zhang, X., Zhang, J., Liu, F., Fu, Z. and Mu, W. (2010). Strengths and limitations on the operating mechanisms of traceability system in agro food, China, *Food Control*. Elsevier 21(6), 825–829.

# Chapter 3

## The role of technology in crisis management and product recall in food supply chains

*Louise Manning, Royal Agricultural University, UK; and Aleksandra Kowalska, Maria Curie-Skłodowska University, Poland*

1 Introduction
2 Food supply chain governance
3 Information asymmetry
4 Data sharing as a mechanism to underpin public-private partnerships (PPPs)
5 Concluding thoughts
6 Where to look for further information
7 References

### 1 Introduction

The first general objective of the European Union (EU) food law is to guarantee a high level of protection of human life, health and consumers' (economic) interests. Protection of fair food trade practices is the next priority (Korzycka and Wojciechowski, 2017). Consumer rights in relation to food are enshrined in Article 9 of Regulation (European Commission (EC)) No 178/2002. Current EU labelling legislation intends to help consumers make informed choices, contributing indirectly to effective competition between businesses that are compliant with the law and consistently producing and selling safe, wholesome food and directly to improved consumer health and welfare. The challenge in a complex food supply chain is to increase consumers' trust in both food regulators and the food industry. Media communication also plays an important and complementary role in governance of food safety and food legality (Zhu et al., 2019). Thus, whilst the safety or legality of food in a given country or region may be seen as being the responsibility of the state to enforce, market mechanisms for food supply, underpinned by definitions of societal, business and individual responsibility, also work together to create a governance structure around food. This governance structure is especially tested during product recalls following a food safety- or legality-related incident.

http://dx.doi.org/10.19103/AS.2021.0097.08

Effective food safety management processes are at the core of a food organisation's strategy, and in the event that such controls fail, crisis management protocols must be in place that can be implemented quickly and effectively (Manning, 2007). A food recall is the 'action taken by a food business to remove unsafe food from distribution, sale and consumption' (Kaaviya et al., 2019, p. 209). Competent product recall processes must include timely identification of the food safety or quality problem, clear identification of the batch of product affected or control of the logistics of product return and/or disposal. They also require processes to be in place that embed mechanisms to retain public trust and confidence in the competency of the regulator, the food organisation, the brand and the food product (Manning, 2007; Jianbin and Hooker, 2019). The Food and Drug Administration (FDA, 2020) define a recall strategy as 'a planned specific course of action to be taken in conducting a specific recall, which addresses the depth of recall, need for public warnings, and extent of effectiveness checks for the recall'. The following terms are also defined by the FDA:

- Market withdrawal – a firm's removal or correction of a distributed product which involves a minor violation that would not be subject to legal action by the FDA or which involves no violation, for example, normal stock rotation practices, routine equipment adjustments and repairs, and so on.
- Recall classification – the numerical designation, that is, I, II or III, assigned by the FDA to a particular product recall to indicate the relative degree of health hazard presented by the product being recalled.

  Class I is a situation in which there is a reasonable probability that the use of, or exposure to, a violative product will cause serious adverse health consequences or death.

  Class II is a situation in which use of, or exposure to, a violative product may cause temporary or medically reversible adverse health consequences or where the probability of serious adverse health consequences is remote.

  Class III is a situation in which use of, or exposure to, a violative product is not likely to cause adverse health consequences.

Article 19 of Regulation (EC) No 178/2002 of the European Parliament and of the Council of 28 January 2002 laying down the general principles and requirements of food law, establishing the European Food Safety Authority and laying down procedures in matters of food safety states: 'If a food business operator considers or has reason to believe that a food which it has imported, produced, processed, manufactured or distributed is not in compliance with the food safety requirements, it shall immediately initiate procedures to withdraw the food in question from the market where the food has left the immediate control of that initial food business operator and inform the competent authorities thereof. Where the product may have reached the consumer, the operator shall effectively and accurately inform the consumers of the reason for its withdrawal, and if necessary, recall from consumers products already supplied to them when other measures are not sufficient to achieve a high level of health protection'.

Jianbin and Hooker (2019) identify three indicators that determine the effectiveness of food recalls. These are discovery time (the days from the production incident to the day the recall is initiated), recovery rate (proportion of the recalled material that is recovered) and completion time (the days from starting the recall to the date of completion). Jianbin and Hooker note that discovery time affects recovery rates and completion time, and there is an association between completion time and the recovery rate. Significant and non-targeted food recalls increase food waste and threaten food security (Vågsholm et al., 2020) and thus traceability systems are essential to support product recall systems (Sufiyan et al., 2019). Food safety crises with associated recalls can drive improvements in regulatory and market governance systems in order to protect public health (Snyder, 2016). This highlights the focus of this chapter on the advances in governance and how this frames crisis management and product recalls in food supply chains. The next section of the chapter considers emerging structures of food supply governance. The challenges of information asymmetry are then explored both in terms of normal operation and in the event of a recall.

## 2 Food supply chain governance

Over three decades ago, Lushbough (1980) has rightly stated that it is already too late when the primary system fails and the food business operator confronts a recall or market withdrawal. This means it is necessary to prevent such incidents and plan far ahead for primary system failure. The involvement of entities from both the public and the private sector in food safety governance will improve the effectiveness of such a preventive policy for product recalls and market withdrawals.

Public-private partnerships (PPPs) between regulators and industry can provide a model for food supply chain governance of the provision of safe and legal food. PPPs can leverage the benefits of public goods and services when the responsibility for the delivery of these public goods and services is shared between the state and the private sector. Hybridisation of the kind that forms PPPs is driven through governance innovation (Pies et al., 2020), that is, PPPs can deliver innovative practices and activities. Hybridisation covers the 'who' (public and private actors) of regulation, the 'how' (command-and-control vs. responsive regulation), the 'where' (global, local and intermediate level), and the 'what' (business vs. social regulation) of policy initiatives (Verbruggen and Havinga, 2017). When compared to purely governmental instruments, PPPs are a partnership of the public and the private sector, aimed at realising activities and rendering services commonly provided by the public sector. Based on the Guidelines for Successful Public–Private Partnerships (EC, 2003), the following four principal roles are identified for the private sector in PPP schemes: to provide additional capital, provide alternative management and implementation skills, provide added value to the consumer and the public at large and to provide better identification of needs and optimal use of resources. Effective PPPs will allow the sharing of resources (financial, human, social and physical), ideas and innovation, reduce duplication and reach a wider audience so activities can be undertaken on a larger scale than individual public and private initiatives (Hernandez-Aguado and Zaragoza, 2016). Further, they argue PPPs allow more diverse communication and marketing channels and wider media access, but conversely, there can be a concern over the independence of such approaches where there is an element of industry self-regulation or a misalignment of public and private goals and objectives.

The structure, efficiency and degree of public–private hybridisation of the food governance system is culturally, historically and structurally conditioned, limiting the ability of food markets to develop and implement universal official food control system instruments across national boundaries where regulations may vary from country to country in terms of what is or is not a legal requirement. To address this challenge and also to ensure that an organisational food governance culture can be established, transnational corporations have developed common standards across their operating markets especially in countries that lack a state-driven food governance system. Indeed, this can lead to transnational corporations operating across multiple governance landscapes with multiple models of food public–private governance depending on the location.

Different PPP governance models co-exist including pure public regulation, and at the other end of the spectrum private organisational and supply chain self-regulation, and a range of hybrid PPPs in between with the resultant 'normative implications' and with the drive for developing PPPs often based on delivering lower regulatory cost (Martinez et al., 2007; Chatzopoulou, 2015). PPP models have been suggested in the academic literature for

- assuring food safety and traceability (Rouviere and Royer, 2017; Narrod et al., 2018), public health and nutrition (Clapp and Scrinis, 2017; Greenberg, 2017; Ibrahim, 2018; Gressier et al., 2020);

- ensuring food security (Zhong et al., 2019), healthy lifestyles and food-related health promotion (Knai et al., 2015; Hernandez-Aguado and Zaragoza, 2016; Seeuws, 2017; De Pinho Campos et al., 2019);
- mitigating food fraud (Spink et al., 2016); and
- creating opportunities for innovation in agricultural systems (Kolaj et al., 2019), cold food chain management (Kendurkar and Tiwari, 2017), agribusiness sustainability (Obayelu, 2018), water allocation and irrigation (Khadra and Sagardoy, 2019; Rafaat et al., 2020), technology access (Mangeni, 2019) and biotechnology (Lokko et al., 2018).

There are degrees of hybridisation in these proposed models for PPPs and tensions can operate within the structures (Manganelli et al., 2019) including corporate power dynamics (Clapp and Scrinis, 2017; Greenberg, 2017), and especially in food supply chain governance (Kowalska and Manning, 2021).

Non-market drivers, such as regulations and changes to regulatory inspection regimes, can affect supply and demand dynamics and influence supply chain interactions. Market failure occurs when markets fail to produce 'either economically optimal [efficient] or socially desirable [equitable] outcomes' (Wolf, 1987, p. 46), such as a food safety incident. For a food supply chain to be socially optimal, that is, the market can consistently supply safe and legal food, it must firstly deliver in practice to consumers and address any negative externalities such as environmental impact and pollution, packaging use and disposal. However, the question arises: what is reasonable to expect, in terms of the information a food organisation is willing to share with others, both in normal times and in the event of a product recall? It is this understanding of what constitutes competitively important and private information that is important here. Differentiating between information organisations would not reasonably be expected to share, often commercially sensitive and providing integral, intangible value in terms of brand equity, and alternatively what information can be reasonably expected shared to promote supply chain and market transparency especially during a product recall is essential. Shen et al. (2019, p. 4898) state that 'information asymmetry refers to the scenario in which some information (e.g. cost information, demand information, supply information, etc.) is considered private and as a result is not publically available to all supply chain members'. Information asymmetry is of particular interest in food supply chain governance approaches and is now considered more specifically.

## 3 Information asymmetry

Knowledge may be positioned within information or may be held as tacit knowledge or know-how. The inability to convert tacit knowledge to explicit knowledge leads to knowledge suppression, and this can cause gaps in knowledge sharing (Teece, 2000). Information asymmetry is simply where one party is in possession of more information than the other, and knowledge asymmetry increases information asymmetry (Nestorowicz, 2014). Information asymmetry occurs 'when an imbalance of knowledge exists between two parties, such as a buyer and a seller, a regulator and an operator, and an employer and an employee' (Salhi, 2020, p. 2983). When sellers have more knowledge than buyers about the integrity of the product, the level of supply chain monitoring and verification associated with products and production processes or how characteristics such as food safety, product quality, provenance, or traceability are assured, then this can influence the power dynamics in their relationship. It is evident from their contractual position in the supply chain that the retailer or food service organisation can collect more consumer and market demand information, whereas the supplier can provide more supply-related information. These two types of information are not always shared openly between all supply chain members. Consumers are the weakest market participants as they are least able to collect and access adequate information. Indeed, consumers bear considerable health and economic risk when purchasing and consuming food (Ozimek, 2012); that is, they can be vulnerable to being misled by food business operators at any stage of the supply chain. Factors such as resource scarcity, property rights, control of market access, criticality of the exchange relationship, incomplete contracts that lead to information asymmetry, asymmetric lock in, adverse selection and morale hazard all play a key role in framing market operations (Manning et al., 2017). Wang et al. (2017) argue that 'retail gatekeepers' could take advantage of consumers because of information asymmetry, although this risk is reduced by implementing effective traceability and labelling systems (Kendall et al., 2019a,b). The potential consequences of information asymmetry are relevant for a broad group of stakeholders, including honest market competitors and consumers who may be vulnerable to sharp practice or the illicit activity of others (Spink and Moyer, 2011; Kowalczyk, 2015; Kwasek, 2015; Marvin et al., 2016; van Ruth et al., 2017; van Ruth et al., 2018; Yang et al., 2019).

Information asymmetry provides an incentive for fraud (McCluskey, 2000; Kowalska et al., 2018), and this concern has been linked to organic food (McCluskey, 2000; Giannakas, 2002) and higher animal welfare products (Bitzios et al., 2017). In the organic food supply chain, information asymmetry exists between farmers and processors, processors and retailers and then retailers and consumers (Manning and Monaghan, 2019). In the organic food market, third-party certification (TPC) plays a vital role in reducing information asymmetry and increasing signalling. TPC verifies the process of production, the credibility of organic producers and releases that information to the public through the issuance of a certificate of conformity and organic food labelling cues. TPC might enhance consumers' trust in the integrity of organic food (Manning and Monaghan, 2019). Consumers' trust is an important determinant of purchase of a credence good, like organic food. Based on TPC as an element of food supply chain governance, long-term contracts between suppliers and retailers can further reduce information asymmetry through both parties participating in the normative definition of production methods, especially where, as in the case of organic food the use of production inputs is controlled (Zhao et al., 2020). In recent times, there has been a dynamic development in a multitude of sustainability-related food labels aimed at reducing information asymmetry between food producers, other businesses in the supply chain and consumers, especially regarding the sustainability impact on the food supply chain (Asioli et al., 2020). Organic food production is just one example of these standards. TPC often provides assurance that sustainable cues on packaging are associated with a set of normative private standards that are routinely verified at steps in the supply chain to ensure integrity of use (Rees et al., 2019).

Supply chain contracts can contribute to overcoming information asymmetry within the supply chain. There are several applications which show that supply chain contracting has been used for solving industrial problems with access to updated information, for example, quick response codes; advance selling; vendor-managed inventory; collaborative planning, forecasting, and replenishment; technological advancement for information acquisition; risk management and sustainable supply chains with demand information updating (Shen et al., 2019).

Market failure occurs in food supply chains when there is information asymmetry between individual actors in the chain (Verbeke, 2005; Manning, 2018). Over two decades ago, Ritson and Mai (1998) state that a market economy might 'fail' in providing safe food because there is asymmetry in risk knowledge; there are aspects of food safety that are perceived as public goods and the social costs of failures in food safety are not borne by the industry and there is a divergence between objective scientific evidence and more subjective consumer perceptions of the risks associated with food safety. Information asymmetry can therefore lead to food safety and quality incidents (Louw and van der Merwe, 2020), a motivation for embedding traceability systems (Hobbs, 2004). Despite new forms of information management systems spreading rapidly, the anonymity of those operating in the food chain is a market challenge, and as previously stated, protecting brand value and fierce price competition generates an incentive and pressure to reduce both knowledge sharing and production costs. In summary, PPPs focus on the co-delivery of food supply chain governance through new and emergent institutional arrangements, whereby the arrangements can be either *consultative* (industry being consulted by the regulator) or *contributory* with mutually beneficial collaboration based on trust, shared goals and accountability mechanisms (Rouviere and Royer,

2017). One opportunity for effective cooperation is the ability to share data between public and private actors, and emerging technologies can support this interaction and access to data and information. The next section of the chapter considers mechanisms for sharing data and one option in particular distributed ledger technologies (DLTs).

## 4 Data sharing as a mechanism to underpin public-private partnerships (PPPs)

A product recall can be very expensive and if badly executed can have a detrimental impact on an organisation's reputation, so effective traceability is a key mechanism to reduce liability (Dai et al., 2015). Recalls therefore use traceability information and knowledge of the product's life cycle (Diallo et al., 2014), and these data can be stored in paper, in digital files or in more sophisticated data systems. Granularity is a major consideration here. Each organisation decides the level of granularity in terms of the level of information precision required (Islam and Cullen, 2021). They argue that:

> "Fine granularity decreases the [traceable resource units] TRU size, but increases overall numbers of TRUs thereby increasing information recording. On the other hand, coarse granularity reduces the level of information recording but the resulting TRU size can obscure important product information."
>
> *(Islam and Cullen, 2021, p. 8)*

Traceability in food supply chains therefore requires an information trail that aligns with the products' physical movements through the supply chain (Smith et al., 2005). Salampasis et al. (2012) consider whether TraceALL, a semantic web ontology-based framework, would provide an infrastructure to enable traceability applications that allow for an information trail associated with the physical trail. They argue that such systems must be 'cost effective, easy to manage and applicable within a globalised, networked, interoperable economic environment' (Salampasis et al., 2012, p. 302). Whilst traceability data can be collated within software systems, it can become outdated as knowledge artefacts lose value. Mills et al. (2018) developed an automatic traceability maintenance system that incorporated machine learning. Blockchain is suggested as a means to ensure that information trails are recorded and data integrity is maintained (Agrawal et al., 2018).

A blockchain quite simply is 'a distributed database, which logs an evolving list of transaction records by organising them into a hierarchical chain of blocks' (Zhang et al., 2019; Creydt and Fischer, 2019). Further, they determine that a blockchain is developed and maintained using a 'peer to peer overlay network' which gains its level of security via 'intelligent and decentralised utilisation of cryptography with crowd computing'. Blockchain technology has been proposed as being of value in assuring product data traceability (Mattila et al., 2016; Kshetri, 2018; Behnke and Janssen, 2019; George et al., 2019; Pearson et al., 2019) and trust in the integrity and confidentiality of information (Kshetri, 2018; Behnke and Janssen, 2019). Therefore, blockchain technology has the potential to not only provide a platform to track and trace product but also to go further to demonstrate provenance via 'providing a robust system to trace origin, certifying authenticity, tracking custody, and verifying the integrity of products' (Montecchi et al., 2019, p. 284). This reduces the potential for identity loss and a failure of traceability systems or, more extremely, instances of food fraud. Since there may be many supply chain members in DLT-based information systems (first tier, second tier, third tier, etc.) who check the information that they have received, they can raise the alarm immediately if they find the information is incorrect or suggests that the product related to the information may be unsafe or out of specification. Therefore, a successful early warning system for identification of potential food safety issues that may develop into a crisis might be blockchain based (Tian, 2018). Implementing blockchain-based solutions should contribute to ensuring quicker recalls along the food supply chain.

DTLs with their transparent transactional activity and data retention are proposed to be of value as a digital form of PPP (Tripoli and Schmidhuber, 2018). DTLs, such as blockchain, 'can provide a cryptographically secure and immutable record of transactions and associated metadata (origin, contracts, process steps, environmental variations, microbial records, etc.) linked across whole supply chains' (Pearson et al., 2019, p. 145).

A blockchain is composed of the following three elements: the block, the chain and the network. The block is composed of a header (a time stamp for when the block was written), a body that records the transaction information and the number of transactions which cannot be modified and a 'hash' which is a link to the previous block (Creydt and Fischer, 2019; Wang et al., 2020). Each block is itself protected by cryptography techniques to enable the trusted integrity of the transaction records (Zhang et al., 2019). The 'hash value' is thus a cryptographic image that defines this block and the previous block in the chain using a hash pointer. This hash of stored data can be verified so data tampering is ultimately recognisable because there is a single root hash pointer to the initial or genesis block (Zhang et al., 2019). The lowest level of the decentralised infrastructure is the signed transaction between two participants, where one or both sign the transaction (Casino et al., 2019). In a transaction between two or multiple parties, the owner of the information retains half of the digital signature (the private key), whilst the other half of the digital signature (the public key) is published to the agreed parties in the network (Cai and Zhu, 2016). However, if the private key of a given party is stolen, it can be very difficult to identify the perpetrator (Efanov and Roschin, 2018). Thus, with appropriate safeguards, blockchain technologies provide the opportunity to identify in a supply chain what data-associated actions have been performed, when and in what location allowing for greater auditability of data and transactions by different supply chain actors (Kshetri, 2017; 2018). Without an auditable data trail, there is a lack of transparency in terms of data integrity from the 'point of generation to the point of data use'; conversely, an auditable trail via blockchain can deliver visibility and accountability (Kshetri, 2017). The use of DLTs reduces information asymmetry and associated power dynamics as the data are distributed, and each actor can access and verify transaction records when needed (Montecchi et al., 2019, p. 284). Whilst private blockchains have limited access, public blockchains are open and subject to privacy issues; the third type, a consortium or federated blockchain, is a mixture of private and public blockchains (Casino et al., 2019; Creydt and Fischer, 2019; Zhang et al., 2019). It is this latter type of chain that could prove to be of value in developing PPPs. Table 1 compares the characteristics of the three types of blockchains. Key characteristics to consider are the level of anonymity, the time taken for the transactions to occur and the level of immutability. The benefit of blockchain is real-time transactions and this technology facilitates auditability for those who have access to the transactions and reduced transaction costs (Kamble et al., 2020). In one study, the use of

**Table 1** Definition and main attributes of different blockchain networks (adapted from Casino et al., 2019; Zhang et al., 2019)

| | Blockchain type | | |
|---|---|---|---|
| Attribute | Public<br>Anyone can participate and is fully accessible and readable | Private<br>Access, read and write permissions are strictly controlled by an organisation | Federated or Consortium<br>Anyone can access and read but only certain participants can have write permissions |
| Consensus | Costly proof of work (PoW) | Light PoW | |
| Ownership and management | Public, permissionless | Centralised, permissioned access | Semi-centralised permissioned nodes |
| Number of trust authorities | 0 | 1 | ≥1 |
| Mechanism | All miners | Centralised organisations | Leader node set |
| Identity | (Pseudo) Anonymous | Identified users (writers) | |
| Anonymity | Opaque | Trusted | |
| Consensus protocol efficiency | Low (slow) | Highest (fastest) | Higher (faster) |
| Transaction approval time | Minutes | Milliseconds | |
| Immutability | Almost impossible | Collusion attacks | |

blockchain technology reduced the time to trace mangoes from 7 days to 2.2 days (Yiannas, 2018).

Blockchain technology can be linked to internet of things (IoT) technology, which is discussed in detail in many chapters of this book. IoT applications can be connected and automatically capture data from sources such as global positioning systems (GPS), geographic information systems, wireless sensor networks and radio frequency identification tags and other barcode systems (Duan et al., 2020). Linking data with GPS can act as a deterrent against theft, and also in the event of theft occurring, materials have the potential to be traced (Bell et al., 2018) or wider provenance (Wallace and Manning, 2020). Others suggest that artificial intelligence applications can support the blockchain-enabled intelligent IoT architecture (Singh et al., 2020). Thus, DLTs can secure 'the evidence chain' (Pearson et al., 2019) especially via smart contracts. Smart contracts are a specific kind of computer programme that works on all nodes in the network where all active users can create a contract by placing a transaction in the system (Tian, 2017). Smart contracts can be more error-free than manual systems and will automatically 'self-execute when specified conditions are met,' reducing transaction costs and making the system more 'real time' (Nathan and Jacobs, 2020). Encrypted unique keys provide a reassurance of data integrity, as when tested the data will generate the same key excepting if the data have been amended (Pearson et al., 2019).

Smart contracts in the event of system or product failure can trigger warnings to prevent a food safety incident, minimising the need for product recall, and can benefit from not needing human intervention (Duan et al., 2020; Keogh et al., 2020). Wu and Lin (2019) propose that the architecture of the system is on multiple layers that are important in the event of a product recall. The blockchain level including smart contracts where information is collected; the process level where data are read, any issues are detected and necessary actions are taken and then the request and validate layer where supply chain actors and regulators and third parties interact with the system. The application of GS1 standards together with blockchain technology can provide for better regulatory control of the use of the technology and enhanced traceability (Keogh et al., 2020).

Challenges for applying blockchain technologies are the variations in laws, customs and practice over national and international jurisdictions, the timescale required to agree DLT-based solutions and the technology adoption in itself may be a barrier to entry for some organisations along with the need to deliver robust cybersecurity measures (Kshetri, 2018). Indeed, in a data-rich supply chain where DLTs are used by some actors and not others, some excluded organisations may be severely disadvantaged by being resource, data and information poor (Arunachalam et al., 2018). Reyna et al. (2018) cite further challenges including anonymity and data privacy versus transparency, creation of consensus, legal issues and security (weaknesses and threats) issues around smart contracts and storage capacity and scalability. Much of this work is still in its infancy, but there is a great potential to reduce information asymmetry and reduce the discovery time and completion time and improve the recovery rate.

## 5 Concluding thoughts

An advantage of blockchain systems within PPPs for food supply chain governance is that verification of transactions can be real time making it more difficult to backdate transactions (Wang and Kogan, 2018), a key need in the event of a food safety incident where a food recall may be required. PPPs for supply chain governance can deliver value at the supply chain and wider stakeholder level including developing blockchain consortia to improve overall efficiency and integrity in information sharing. However, ownership rights and allocation of permissions, competitive issues that may create challenges for private business partners, identifying which actors are the guarantors of data quality and the ethical use of data that are collected and stored in the system all need to be considered (Schwabe, 2019). This is addressed by the regulator or an independent organisation appointed by the regulator often termed a data trust. For such PPPs, based on a blockchain system, to be effective in practice, operating measures need to be in place that address membership criteria, system architecture, data management protocols and governance structures (Nathan and Jacobs, 2020).

Some actors may resist information sharing if they consider that they will impinge on competitive advantage or they have a modus operandi that focuses on information asymmetry as a means to drive personal gain (Chadderton and Norton, 2019). However, Chadderton and Norton argue that when implementing PPPs they should focus not only on legislative and technological challenges but also on factors such as relationships and organisational culture, trust and differentiated risk appetite through long-term commitment, effective governance, collectively agreed strategic direction, shared resources and clear processes to manage data. Further, they conclude that collaboration between government and the private sector through the creation of PPPs deepens individual and collective understanding of dark, opaque behaviour and as a result drives the development of effective mitigation strategies.

Blockchain technologies can improve information flow within PPPs as each node in the network can be related to a given IP address and an 'event log' with transactions between actors being verified to ensure integrity as they become immutable after being submitted to the system (Hyvärinen et al., 2017); that is, they are tamper resistant (Beck et al., 2017). The immutability of the logs of past transactions means that the system can be verified by multiple actors

(Hyvärinen et al., 2017; Casado-Vara et al., 2018), even if they do not trust each other (Wang and Kogan, 2018; Bell et al., 2018). Beck et al. (2017, p. 382) state that blockchain could overcome information asymmetry by providing 'a decentralised global information infrastructure in which no one is in full control, no one has absolute power, and no one can distort or lie about past or current events'. Information sent to the blockchain can never be erased or changed. This is of particular value in maintaining PPPs; however, access to immutable information can only be achieved in a public or federated blockchain as it is possible to manipulate private blockchains if the central authority is corrupt, thus there is a 'trade-off between information confidentiality [privacy] and transparency' (Wang and Kogan, 2018, p. 2).

Information asymmetry is a key barrier to supply chain development, innovation and efficiency and effective product recalls. Food supply chain governance PPPs provide an opportunity to improve transparency to ensure effective open regulation, an understanding of market standards and the opportunities for market access and a reduction in supply chain opacity. PPPs and a range of hybrid models can be operationalised suited to a given regulatory and market situation, especially a product recall. Further work needs to be undertaken to frame such approaches, including who the members of such PPPs should be and how the PPPs are operationalised in practice. In March 2021, The UK Food Standards Agency published a research report Food Data Trust: A Framework for Information Sharing, and the report highlights that easier data sharing could 'speed up information exchange along a chain in urgent situations such as food recalls and tracing incidents' (p. 2). Good governance around data collection, sharing and storage is essential. The design and implementation of structures that can deliver good governance of food supply chains and including engagement with associated regulatory bodies, especially during a product recall, is in its infancy as we write this chapter. However, this is an exciting area of emergent research and policy, and the outputs in terms of policy tools and governance structures should serve the industry and consumers in the future.

## 6 Where to look for further information

### *6.1 Public Private Partnership*

- Public Private Partnership. Available at: https://ec.europa.eu/regional_policy/sources/docgener/guides/ppp/intro_fiche.pdf.
- Public Private Partnerships with the food industry. Available at: https://www.paho.org/hq/dmdocuments/2015/ppptg-discussion-paper.PDF.

### 6.2 Blockchain, supply chain standards and traceability

- Keogh, J. G., Rejeb, A., Khan, N., Dean, K. and Hand, K. J. (2020). Blockchain and GS1 standards in the food chain: a review of the possibilities and challenges. In: Building the Future of Food Safety Technology (1st Edition). London, UK: Oxford.

### 6.3 Data governance and trust frameworks

- Brewer, S., Pearson, S., Maull, R., Godsiff, P., Frey, J. G., Zisman, A., Parr, G., McMillan, A., Cameron, S., Blackmore, H. and Manning, L. (2021). A trust framework for digital food systems. Nature Food 2(8), 543–545.

## 7 References

Agrawal, T. K., Sharma, A. and Kumar, V. (2018). Blockchain-based secured traceability system for textile and clothing supply chain. In: *Artificial Intelligence for Fashion Industry in the Big Data Era* Thomassey, S. and Zeng, X. (Eds). Springer: Singapore (pp. 197–208).

Arunachalam, D., Kumar, N. and Kawalek, J. P. (2018). Understanding big data analytics capabilities in supply chain management: unravelling the issues, challenges and implications for practice. *Transportation Research Part E: Logistics and Transportation Review* 114, 416–436. https://doi.org/10.1016/j.tre.2017.04.001.

Asioli, D., Aschemann-Witzel, J. and Nayga Jr., R. M. (2020). Sustainability-related food labels. *Annual Review of Resource Economics* 12(1), 171–185. https://doi.org/10.1146/annurev-resource-100518-094103.

Beck, R., Avital, M., Rossi, M. and Thatcher, J. B. (2017). Blockchain technology in business and information systems research. *Business and Information Systems Engineering* 59(6), 381–384. https://doi.org/10.1007/s12599-017-0505-1.

Behnke, K. and Janssen, M. F. W. H. A. (2019). Boundary conditions for traceability in food supply chains using blockchain technology. *International Journal of Information Management* 52. https://doi.org/10.1016/j.ijinfomgt.2019.05.025.

Bell, L., Buchanan, W. J., Cameron, J. and Lo, O. (2018). Applications of blockchain within healthcare. *Blockchain in Healthcare Today* 1, 1–7. https://doi.org/10.30953/bhty.v1.8.

Bitzios, M., Jack, L., Krzyzaniak, S. C. and Xu, M. (2017). Chapter 20. Dissonance in the food traceability regulatory environment and food fraud. In: *It's a Jungle Out There-the Strange Animals of Economic Organization in agri-Food Value Chains* Martino, G., et al. (Eds). Wageningen Academic Publishers: Wageningen (pp. 375–393). https://doi.org/10.3920/978-90-8686-844-5_20.

Cai, Y. and Zhu, D. (2016). Fraud detections for online businesses: a perspective from blockchain technology. *Financial Innovation* 2(1), 1–10. https://doi.org/10.1186/s40854-016-0039-4.

Casado-Vara, R., Prieto, J. and Corchado, J. M. (2018). How blockchain could improve fraud detection in power distribution grid. In: *Advances in Intelligent Systems and Computing The 13th International Conference on Soft Computing Models in*

*Industrial and Environmental Applications*. Springer: Cham (pp. 67–76). https://doi.org/10.1007/978-3-319-94120-2_7.

Casino, F., Dasaklis, T. K. and Patsakis, C. (2019). A systematic literature review of blockchain-based applications: current status, classification and open issues. *Telematics and Informatics* 36, 55–81. https://doi.org/10.1016/j.tele.2018.11.006.

Chadderton, P. and Norton, S. (2019). Public-private partnerships to disrupt financial crime: an exploratory study of Australia's Fintel Alliance. *SSRN Electronic Journal*. SWIFT Institute Working Paper No. 2019-003. Available at: https://ssrn.com/abstract=3392268 or http://dx.doi.org/10.2139/ssrn.3392268.

Chatzopoulou, S. (2015). The dynamics of the transnational food chain regulatory governance: an analytical framework. *British Food Journal* 117(10), 2609–2627. https://doi.org/10.1108/BFJ-11-2014-0368.

Clapp, J. and Scrinis, G. (2017). Big food, nutritionism, and corporate power. *Globalizations* 14(4), 578–595. https://doi.org/10.1080/14747731.2016.1239806.

Creydt, M. and Fischer, M. (2019). Blockchain and more-algorithm driven food traceability. *Food Control* 105, 45–51. https://doi.org/10.1016/j.foodcont.2019.05.019.

Dai, H., Tseng, M. M. and Zipkin, P. H. (2015). Design of traceability systems for product recall. *International Journal of Production Research* 53(2), 511–531.

De Pinho Campos, K., Cohen, J. E., Gastaldo, D. and Jadad, A. R. (2019). Public-private partnership (PPP) development: toward building a PPP framework for healthy eating. *International Journal of Health Planning and Management* 34(1), e142–e156. https://doi.org/10.1002/hpm.2714.

Diallo, T. M., Henry, S. and Ouzrout, Y. (2014). Using unitary traceability for an optimal product recall. In: *IFIP International Conference on Advances in Production Management Systems*. Springer: Berlin, Heidelberg (pp. 159–166).

Duan, J., Zhang, C., Gong, Y., Brown, S. and Li, Z. (2020). A content-analysis based literature review in blockchain adoption within food supply chain. *International Journal of Environmental Research and Public Health* 17(5), 1784.

EC (European Commission) (2003). Guidelines for successful public-private partnerships. Available at: https://ec.europa.eu/regional_policy/sources/docgener/guides/ppp_en.pdf (Accessed 4 October 2019).

Efanov, D. and Roschin, P. (2018). The all-pervasiveness of the blockchain technology. *Procedia Computer Science* 123, 116–121. https://doi.org/10.1016/j.procs.2018.01.019.

FDA (Food Drug Administration) (2020). Recalls, corrections and removals (devices). Available at: https://www.fda.gov/medical-devices/postmarket-requirements-devices/recalls-corrections-and-removals-devices (Accessed 30 May 2021).

Food Standards Agency (2021). Food data trust: a framework for information sharing. Available at: https://www.food.gov.uk/sites/default/files/media/document/food-data-trust-a-framework-for-information-sharing.pdf.

George, R. V., Harsh, H. O., Ray, P. and Babu, A. K. (2019). Food quality traceability prototype for restaurants using blockchain and food quality data index. *Journal of Cleaner Production* 240, 118021. https://doi.org/10.1016/j.jclepro.2019.118021.

Giannakas, K. (2002). Information asymmetries and consumption decisions in organic food product markets. *Canadian Journal of Agricultural Economics/Revue Canadienne d&Apos;Agroeconomie* 50(1), 35–50. https://doi.org/10.1111/j.1744-7976.2002.tb00380.x.

Greenberg, S. (2017). Corporate power in the agro-food system and the consumer food environment in South Africa. *The Journal of Peasant Studies* 44(2), 467–496.

Gressier, M., Sassi, F. and Frost, G. (2020). Healthy foods and healthy diets. How government policies can steer food reformulation. *Nutrients* 12(7), 1992. https://doi.org/10.3390/nu12071992.

Hernandez-Aguado, I. and Zaragoza, G. A. (2016). Support of public–private partnerships in health promotion and conflicts of interest. *BMJ Open* 6(4), e009342. http://doi.org/10.1136/bmjopen-2015-009342.

Hobbs, J. E. (2004). Information asymmetry and the role of traceability systems. *Agribusiness* 20(4), 397–415.

Hyvärinen, H., Risius, M. and Friis, G. (2017). A blockchain-based approach towards overcoming financial fraud in public sector services. *Business and Information Systems Engineering* 59(6), 441–456. https://doi.org/10.1007/s12599-017-0502-4.

Ibrahim, U. (2018). Public-private partnerships a panacea to obesity crisis in South East Asia. *Noble International Journal of Social Sciences Research* 3(3), 14–20.

Islam, S. and Cullen, J. M. (2021). Food traceability: a generic theoretical framework. Food Control, 107848. https://doi.org/10.1016/j.foodcont.2020.107848.

Jianbin, Y. and Hooker, N. H. (2019). Exploring relationships among recall effectiveness indicators in the US meat and poultry industry. *International Food and Agribusiness Management Review* 22(1)(1030-2019-620, 97–106.

Kaaviya, C., Lavanya, S. M. and Krishnakumare, B. (2019). Consumers' opinion towards food product recall. *European Journal of Nutrition and Food Safety*, 208–215. Available at: https://journalejnfs.com/index.php/EJNFS/article/view/30114.

Kamble, S. S., Gunasekaran, A. and Sharma, R. (2020). Modeling the blockchain enabled traceability in agriculture supply chain. *International Journal of Information Management* 52, 101967.

Kendall, H., Clark, B., Rhymer, C., Kuznesof, S., Hajslova, J., Tomaniova, M., Brereton, P. and Frewer, L. (2019a). A systematic review of consumer perceptions of food fraud and authenticity: a European perspective. *Trends in Food Science and Technology* 94, 79–90. https://doi.org/10.1016/j.tifs.2019.10.005.

Kendall, H., Kuznesof, S., Dean, M., Chan, M. Y., Clark, B., Home, R., Stolz, H., Zhong, Q., Liu, C., Brereton, P. and Frewer, L. (2019b). Chinese consumer's attitudes, perceptions and behavioural responses towards food fraud. *Food Control* 95, 339–351. https://doi.org/10.1016/j.foodcont.2018.08.006.

Kendurkar, P. and Tiwari, A. (2017). Cold chain supply and public-private partnership: a proactive interaction. *International Journal of Scientific Research in Multidisciplinary Studies* 3(9), 1–6.

Keogh, J. G., Rejeb, A., Khan, N., Dean, K. and Hand, K. J. (2020). Optimising global food supply chains: The case for blockchain and GS1 standards. In: *Building the Future of Food Safety Technology* Blockchain and beyond. Detwiler, D. (Ed.). Elsevier London. ISBN 978-0-12-818956-6 pp.176–196.

Khadra, R. and Sagardoy, J. A. (2019). Private public partnerships (PPPs) in irrigation. Myth or promising reality? In: *Irrigation Governance Challenges in the Mediterranean Region: Learning from Experiences and Promoting Sustainable Performance*. Springer: Cham (pp. 163–199). https://doi.org/ 10.1007/978-3-030-13554-6_6.

Knai, C., Petticrew, M., Durand, M. A., Eastmure, E., James, L., Mehrotra, A., Scott, C. and Mays, N. (2015). Has a public–private partnership resulted in action on healthier diets in England? An analysis of the public health responsibility Deal food pledges. *Food Policy* 54, 1–10. https://doi.org/10.1016/j.foodpol.2015.04.002.

Kolaj, R., Osmani, M., Borisov, P. and Skunca, D. (2019). Empowering partnering links as opportunities for development of the regions: can PPPs work in agriculture. *Bulgarian Journal of Agricultural Science* 25(3), 468–473.

Korzycka, M. and Wojciechowski, P. (2017). *System prawa żywnościowego [The System of Food Law]*. Wolters Kluwer: Warsaw SA.

Kowalczyk, S. (2015). Wzrost turbulencji na rynku globalnym a bezpieczeństwo [Increase in turbulence on the global market vs. safety]. In: *Przeciwdziałanie Turbulencjom w Gospodarce [Counteracting Turbulence in the Economy]* Sobiecki, R. (Ed.). Oficyna Wydawnicza SGH: Warsaw (pp. 2–64).

Kowalska, A. and Manning, L. (2021). Using the rapid alert system for food and feed: potential benefits and problems on data interpretation. Critical Reviews in Food Science and Nutrition 61(6), 906–919. https://doi.org/10.1080/10408398.2020.1747978.

Kowalska, A., Soon, J. M. and Manning, L. (2018). A study on adulteration in cereals and bakery products from Poland including a review of definitions. *Food Control* 92, 348–356. https://doi.org/10.1016/j.foodcont.2018.05.007.

Kshetri, N. (2017). Blockchain's roles in strengthening cybersecurity and protecting privacy. *Telecommunications Policy* 41(10), 1027–1038. https://doi.org/10.1016/j.telpol.2017.09.003.

Kshetri, N. (2018). 1 Blockchain's roles in meeting key supply chain management objectives. *International Journal of Information Management* 39, 80–89. https://doi.org/10.1016/j.ijinfomgt.2017.12.005.

Kwasek, M. (Ed.) (2015). *Z badań nad rolnictwem społecznie zrównoważonym (33) Analiza bezpieczeństwa żywnościowego Polski [From research on socially sustainable agriculture* [33] *Food security analysis for Poland]. Monografie Programu Wieloletniego 2015–2019 nr 19*. IERiGŻ PIB: Warsaw.

Lokko, Y., Heijde, M., Schebesta, K., Scholtès, P., Van Montagu, M. and Giacca, M. (2018). Biotechnology and the bioeconomy–towards inclusive and sustainable industrial development. *New Biotechnology* 40(A), 5–10. https://doi.org/10.1016/j.nbt.2017.06.005.

Louw, M. and van der Merwe, M. (2020). Asymmetry in food safety information–the case of the 2018 Listeriosis outbreak and low-income, urban consumers in Gauteng, South Africa. *Agrekon* 59(2), 129–143.

Lushbough, C. H. (1980). When the primary system fails: recalls, market withdrawals and stock recoveries. *Journal of Food Quality* 3(2), 97–102. https://doi.org/10.1111/j.1745-4557.1980.tb00694.x.

Manganelli, A., Van den Broeck, P. and Moulaert, F. (2019). Socio-political dynamics of alternative food networks: a hybrid governance approach. *Territory, Politics, Governance* 8(3), 299–318. https://doi.org/10.1080/21622671.2019.1581081.

Mangeni, B. (2019). The role of Public-Private Partnerships (PPPs) in ensuring technology access for farmers in sub-Saharan Africa. *African Journal of Food, Agriculture, Nutrition and Development* 19(1), 14137–14155. https://doi.org/10.18697/ajfand.84.BLFB1018.

Manning, L. (2007). Food safety and brand equity. *British Food Journal* 109(7), 496–510. https://doi.org/10.1108/00070700710761491.

Manning, L. (2018). Food supply chain fraud: the economic environmental and socio-political consequences. In: *Advances in Food Security and Sustainability* (vol. 3) Barling, D. and Fanzo, J. (Eds). Academic Press: San Diego, CA. https://doi.org/10.1016/bs.af2s.2018.09.001.

Manning, L. and Monaghan, J. (2019). Integrity in the fresh produce supply chain: solutions and approaches to an emerging issue. *Journal of Horticultural Science and Biotechnology* 94(4), 413–421 https://doi.org/10.1080/14620316.2019.1574613.

Manning, L., Soon, J. M., Aguiar, L. K., Eastham, J. F. and Higashi, S. Y. (2017). Pressure: driving illicit behaviour in the food supply chain 12th Research Workshop on Institutions and Organisations (12th RWIO) Brazil 10.

Martinez, M. G., Fearne, A., Caswell, J. A. and Henson, S. (2007). Co-regulation as a possible model for food safety governance: opportunities for public-private partnerships. *Food Policy* 32(3), 299–314. https://doi.org/10.1016/j.foodpol.2006.07.005.

Marvin, H. J. P., Bouzembrak, Y., Janssen, E. M., Van der Fels-Klerx, H. J., van Asselt, E. D. and Kleter, G. A. (2016). A holistic approach to food safety risks: food fraud as an example. *Food Research International* 89(1), 463–470. https://doi.org/10.1016/j.foodres.2016.08.028.

Mattila, J., Seppälä, T. and Holmström, J. (2016). Product-centric information management – a case study of a shared platform with blockchain technology. Conference Paper. Industry Studies Association Conference.

McCluskey, J. J. (2000). A game theoretic approach to organic foods: an analysis of asymmetric information and policy. *Agricultural and Resource Economics Review* 29(1), 1–9. https://doi.org/10.1017/S1068280500001386.

Mills, C., Escobar-Avila, J. and Haiduc, S. (2018, September). Automatic traceability maintenance via machine learning classification. In: IEEE International Conference on Software Maintenance and Evolution (ICSME) (vol. 2018) (pp. 369–380). IEEE Publications.

Montecchi, M., Plangger, K. and Etter, M. (2019). It's real, trust me! Establishing supply chain provenance using blockchain. *Business Horizons* 62(3), 283–293. https://doi.org/10.1016/j.bushor.2019.01.008.

Narrod, C., Dou, X., Wychgram, C. and Miller, M. (2018). Economic rationale for US involvement in public-private partnerships in international food safety capacity building. In: *Food Safety Economics*. Springer: Cham (pp. 267–291). https://doi.org/10.1007/978-3-319-92138-9_14.

Nathan, J. and Jacobs, B. (2020). Blockchain consortium networks: adding security and trust in financial services. *Journal of Corporate Accounting and Finance* 31(2), 29–33.

Nestorowicz, R. (2014). The information asymmetry and the social responsibility on the food market. International Journal of Arts & Sciences 7(2), 59.

Obayelu, A. E. (2018). Public-private partnerships for inclusive agribusiness sustainability in Africa. *Agriculturae Conspectus Scientificus* 83(3), 251–261.

Ozimek, I. (2012). Ochrona konsumentów na rynku żywności – wybrane aspekty [Selected aspects of consumer protection on the food market]. *Konsumpcja i Rozwój*, 1, 61–70.

Pearson, S., May, D., Leontidis, G., Swainson, M., Brewer, S., Bidaut, L., Frey, J. G., Parr, G., Maull, R. and Zisman, A. (2019). Are distributed ledger technologies the panacea for food traceability? *Global Food Security* 20, 145–149. https://doi.org/10.1016/j.gfs.2019.02.002.

Pies, I., Hielscher, S. and Everding, S. (2020). Do hybrids impede sustainability? How semantic reorientations and governance reforms can produce and preserve sustainability in sharing business models. *Journal of Business Research* 115, 174–185. https://doi.org/10.1016/j.jbusres.2020.04.024.

Rafaat, R., Osman, H., Georgy, M. and Elsaid, M. (2020). Preferred risk allocation in Egypt's water sector PPPs. *International Journal of Construction Management* 20(6), 585–597. https://doi.org/10.1080/15623599.2019.1703087.

Rees, W., Tremma, O. and Manning, L. (2019). Sustainability cues on packaging: the influence of recognition on purchasing behavior. *Journal of Cleaner Production* 235, 841–853. https://doi.org/10.1016/j.jclepro.2019.06.217.

Regulation EC/178/2002 laying down the general principles and requirements of food safety law, establishing the European Food Standards Agency and laying down procedures in matters of food safety OJ L/31 1.2.2002 pp. 001 – 024.

Reyna, A., Martín, C., Chen, J., Soler, E. and Díaz, M. (2018). On blockchain and its integration with IoT. Challenges and opportunities. *Future Generation Computer Systems* 88, 173–190. https://doi.org/10.1016/j.future.2018.05.046.

Ritson, C. and Mai, L. W. (1998). The economics of food safety. *Nutrition and Food Science* 98(5), 253–259. https://doi.org/10.1108/00346659810224163.

Rouviere, E. and Royer, A. (2017). Public Private Partnerships in food industries: a road to success? *Food Policy* 69, 135–144. https://doi.org/10.1016/j.foodpol.2017.04.003.

Salampasis, M., Tektonidis, D. and Kalogianni, E. P. (2012). TraceALL: a semantic web framework for food traceability systems. *Journal of Systems and Information Technology* 14(4), 302–317. https://doi.org/10.1108/13287261211279053.

Salhi, Y. (2020, April). A framework for measuring information asymmetry. In: *Proceedings of the AAAI Conference on Artificial Intelligence* 34(3), 2983–2990.

Schwabe, G. (2019). The role of public agencies in blockchain consortia: learning from the Cardossier. *Information Polity* 24(4), 437–451. https://doi.org/10.3233/IP-190147.

Seeuws, C. (2017). Belgian branded food products database: inform consumers on a healthy lifestyle in a public-private partnership. *Journal of Food Composition and Analysis* 64, 39–42. https://doi.org/10.1016/j.jfca.2017.07.008.

Shen, B., Choi, T. M. and Minner, S. (2019). A review on supply chain contracting with information considerations: information updating and information asymmetry. International Journal of Production Research 57(15–16), 4898–4936.

Singh, S. K., Rathore, S. and Park, J. H. (2020). Blockiotintelligence: a blockchain-enabled intelligent IoT architecture with artificial intelligence. *Future Generation Computer Systems* 110, 721–743.

Smith, G. C., Tatum, J. D., Belk, K. E., Scanga, J. A., Grandin, T. and Sofos, J. N. (2005). Traceability from a US perspective. *Meat Science* 71(1), 174–193.

Snyder, F. (2016). Three worlds of melamine. In: *Food Safety Law in China* (vol. 6) (pp. 9–107). Brill. https://doi.org/10.1163/9789004306929_003.

Spink, J. and Moyer, D. C. (2011). Defining the public health threat of food fraud. *Journal of Food Science* 76(9), R157–R163. https://doi.org/10.1111/j.1750-3841.2011.02417.x.

Spink, J., Moyer, D. C. and Whelan, P. (2016). The role of the public private partnership in Food Fraud prevention–includes implementing the strategy. *Current Opinion in Food Science* 10, 68–75. https://doi.org/10.1016/j.cofs.2016.10.002.

Sufiyan, M., Haleem, A., Khan, S. and Khan, M. I. (2019). Analysing attributes of food supply chain management: a comparative study. In: *Lecture Notes in Mechanical Engineering*. Springer: Singapore (pp. 515–523).

Teece, D. J. (2000). Strategies for managing knowledge assets: the role of firm structure and industrial context. *Long Range Planning* 33(1), 35–54.

Tian, F. (2017, June). A supply chain traceability system for food safety based on HACCP, blockchain & Internet of things. In: International Conference on Service Systems and Service Management (vol. 2017). IEEE Publications (pp. 1–6). Available at: https://dpo.org/10.1109/ICSSSM.2017.7996119.

Tian, F. (2018). An information system for food safety monitoring in supply chains based on HACCP, blockchain and internet of things. Doctoral thesis. WU Vienna University of Economics and Business. Available at: https://epub.wu.ac.at/6090/ (Accessed 05.12.2019).

Tripoli, M. and Schmidhuber, J. (2018). Emerging opportunities for the application of blockchain in the Agri-food industry. Food and Agriculture Organization and International Center for Trade and Sustainable Development: Rome and Geneva. *Licence: CC BY-NC-SA* (pp. 1–41). Available at: http://www.fao.org/3/CA1335EN/ca1335en.pdf (Accessed 5 July 2020).

Vågsholm, I., Arzoomand, N. S. and Boqvist, S. (2020). Food security, safety, and sustainability - getting the trade-offs right. *Frontiers in Sustainable Food Systems* 4, 16. https://doi.org/10.3389/fsufs.2020.00016.

van Ruth, S. M., Huisman, W. and Luning, P. A. (2017). Food fraud vulnerability and its key factors. *Trends in Food Science and Technology* 67, 70–75. https://doi.org/10.1016/j.tifs.2017.06.017.

van Ruth, S. M., Luning, P. A., Silvis, I. C. J., Yang, Y. and Huisman, W. (2018). Differences in fraud vulnerability in various food supply chains and their tiers. *Food Control* 84, 375–381. https://doi.org/10.1016/j.foodcont.2017.08.020.

Verbeke, W. (2005). Agriculture and the food industry in the information age. *European Review of Agricultural Economics* 32(3), 347–368. https://doi.org/10.1093/eurrag/jbi017.

Verbruggen, P. and Havinga, T. (2017). Hybridization of food governance: an analytical framework. In: *Hybridization of Food Governance. Trends, Types and Results* Verbruggen, P. and Havinga, T. (Eds). Edward Elgar Publishing: Cheltenham, UK (pp. 1–28).

Wallace, C. A. and Manning, L. (2020). Food provenance: assuring product integrity and identity, CAB Reviews, ISSN 1749–8848.

Wang, C. S., Van Fleet, D. D. and Mishra, A. K. (2017). Food integrity: a market-based solution. *British Food Journal* 119(1), 7–19. https://doi.org/10.1108/BFJ-04-2016-0144.

Wang, Q., Zhu, X., Ni, Y., Gu, L. and Zhu, H. (2020). Blockchain for the IoT and industrial IoT: a review. *Internet of Things* 10. https://doi.org/10.1016/j.iot.2019.100081, 100081.

Wang, Y. and Kogan, A. (2018). Designing confidentiality-preserving Blockchain-based transaction processing systems. *International Journal of Accounting Information Systems* 30, 1–18. https://doi.org/10.1016/j.accinf.2018.06.001.

Wolf, C. (1987). Market and non-market failures: comparison and assessment. Journal of Public Policy 7(1), 43–70.

Wu, X. and Lin, Y. (2019). Blockchain recall management in pharmaceutical industry. *Procedia CIRP* 83, 590–595.

Yang, Y., Huisman, W., Hettinga, K. A., Liu, N., Heck, J., Schrijver, G. H., Gaiardoni, L. and van Ruth, S. M. (2019). Fraud vulnerability in the Dutch milk supply chain: assessments of farmers, processors and retailers. *Food Control* 95, 308–317. https://doi.org/10.1016/j.foodcont.2018.08.019.

Yiannas, F. (2018). A new era of food transparency powered by blockchain. *Innovations: Technology, Governance, Globalization* 12(1–2), 46–56.

Zhang, R., Xue, R. and Liu, L. (2019). Security and privacy on blockchain. *ACM Computing Surveys (CSUR)* 52(3), 1–34. https://doi.org/10.1145/3316481.

Zhao, J., Gerasimova, K., Peng, Y. and Sheng, J. (2020). Information asymmetry, third party certification and the integration of organic food value chain in China. *China Agricultural Economic Review* 12(1), 20–38. https://doi.org/10.1108/CAER-05-2018-0111.

Zhong, T., Si, Z., Crush, J., Scott, S. and Huang, X. (2019). Achieving urban food security through a hybrid public-private food provisioning system: the case of Nanjing, China. *Food Security* 11(5), 1071–1086. https://doi.org/10.1007/s12571-019-00961-8.

Zhu, X., Yuelu Huang, I. and Manning, L. (2019). The role of media reporting in food safety governance in China: a dairy case study. *Food Control* 96, 165–179. https://doi.org/10.1016/j.foodcont.2018.08.027.

# Chapter 4

# Improving supply chains to prevent food losses and waste: an overview

*Hamid El Bilali, University of Natural Resources and Life Sciences (BOKU), Austria*

1 Introduction
2 Food losses and waste along the supply chain: main causes and reduction strategies
3 Reducing food wastage along the food supply chain: the need for a systems approach
4 Future trends and conclusion
5 Where to look for further information
6 References

## 1 Introduction

Food losses and waste (FLW) occur along the whole supply chain (harvesting, transport, storage, packaging, processing, wholesale, retail) and where food is consumed. About a third of the global food production for human consumption is lost or wasted. Losses in the first part of the food chain are more important in developing countries, while food wastage at the retail and consumer levels is more relevant in industrialised countries (FAO, 2011; HLPE, 2014).

The reduction of FLW is one of the topics addressed in the '2030 Agenda for Sustainable Development', including the 'Sustainable Development Goals' (SDGs) (United Nations, 2015). The target 12.3 of SDG12 (Ensure sustainable consumption and production patterns) seeks to halve per capita global food waste at the retail and consumer levels and to reduce food losses along production and supply chains, including post-harvest losses. Bengtsson et al. (2018) argue that the SDG target 12.3 is also aiming for enhanced efficiency of agro-food systems.

FLW are undermining the very foundations of food security. Therefore, the reduction of the amount of food lost or wasted is now presented as essential to improve food security (FAO, 2012a,b,c, 2017; HLPE, 2011, 2014; Smil, 2004). The amount of food produced on farmers' fields is much more than is necessary to feed humanity. Nevertheless, losses of food between the farmer's field and the dinner table are huge (Corrado and Sala, 2018; FAO, 2011; HLPE, 2014;

http://dx.doi.org/10.19103/AS.2019.0053.08

Parfitt et al., 2010; Smil, 2004; Stuart, 2009). When converted into calories, global FLW amount to approximately 24% of all food produced (Kummu et al., 2012; Lipinski et al., 2016; Searchinger et al., 2013). Even if just one-fourth of the food currently lost or wasted globally could be saved, it would be enough to feed 870 million hungry people (FAO, 2013). Similarly, FAO, IFAD, UNICEF, WFP and WHO (2017) report more recently that all the world's 815 million hungry people could be lifted out of energy/protein malnourishment on less than a quarter of food wasted in the United States and Europe. Nevertheless, reduction in food wastage[1] would improve food and nutrition security (FNS) only if also the food distribution problem is tackled (HLPE, 2014).

FLW are undermining the sustainability of the current food consumption and production patterns. Therefore, FLW reduction constitutes a significant lever to achieve food security and improve food chain efficiency (e.g. FAO, 2011). Reducing waste across the whole food chain would improve input use efficiency by increasing the amount of available food for a given level of inputs (Alexander et al., 2017; FAO, 2013, 2017; Ingram, 2011; Scherhaufer et al., 2018; Smil, 2004; TEEB, 2018; Usubiaga et al., 2018; Wunderlich and Martinez, 2018). In fact, reducing food wastage would ease the pressure on agro-food systems to meet the growing food demand thus improving the sustainability of food systems while contributing to food security.

This chapter explores the main causes of FLW and provides an overview on strategies to prevent and/or to reduce them along the supply chain.

## 2 Food losses and waste along the supply chain: main causes and reduction strategies

The nature, causes and extent of food wastage differ among developed economies, developing countries and economies in transition (Alamar et al., 2018; FAO, 2011; HLPE, 2014). In developing countries, about 40% of losses occurs at post-harvest and processing levels while in industrialised countries more than 40% of losses happens at retail and consumer levels (HLPE, 2014). Poverty and/or limited household income in developing countries make it unacceptable to waste food (FAO, 2011). The exact causes of food losses depend on the local situation in a given country. Generally speaking, FLW are influenced by agricultural production patterns, domestic infrastructure, marketing chains and distribution channels, and consumer food-related

1 Unless indicated otherwise, the term 'food wastage' refers in this chapter to both 'food losses' and 'food waste': Food wastage refers to any food lost by deterioration or waste. Thus, the term 'wastage' encompasses both food loss and food waste (FAO, 2013:9). Furthermore, for the purpose of this chapter, reference is made to the following definitions of the High Level Panel of Experts on Food Security and Nutrition of the Committee on World Food Security (HLPE, 2014:22): Food loss and waste (FLW) refers to a decrease, at all stages of the food chain from harvest to consumption, in mass, of food that was originally intended for human consumption, regardless of the cause. Food losses (FL) refers to a decrease, at all stages of the food chain prior to the consumer level, in mass, of food that was originally intended for human consumption, regardless of the cause. Food waste (FW) refers to food appropriate for human consumption being discarded or left to spoil at consumer level - regardless of the cause.

practices (FAO, 2011). Likewise, solutions for reducing FLW differ across countries and agricultural sub-sectors (crop production, animal production, fisheries). Moreover, the potential for FLW reduction differs along the food chain. Kummu et al. (2012) argue that the largest global potential for improvement is in agricultural losses and consumption waste; global agricultural losses could be reduced by 47% and global consumption waste by 86%. Potential solutions to FLW reduction are also quite different across countries. Improving food supply chain efficiency seems to be the key for developing countries (Kader, 2005), while developed countries should conduct consumer education campaigns and facilitate increased surplus food donation (e.g. food banks) (European Commission, 2010). Innovation (product, process, organisational, social) along the food chain is crucial in the reduction of FLW (e.g. Di Terlizzi et al., 2016; Porat et al., 2018).

Many food waste-management hierarchies or pyramids have been developed in the last few years. These prioritise wastage reduction at source and present a list of preferences for use, re-use, recycling and waste treatment. In general, all food-use hierarchies prioritise FLW prevention and food redistribution of, still edible but not marketable, food, for example by means of food banks[2] (HLPE, 2014). In fact, they prioritise these two options with respect to using FLW as animal feed, compost and/or energy or disposal in landfills, which is considered the last recourse and least preferred option (HLPE, 2014). This chapter focuses on options regarding reduction of FLW through prevention and/or redistribution.

## *2.1 Agricultural production and harvesting*

In developing countries, production and post-harvest handling contribute the greater shares of the total food losses (e.g. Aragie et al., 2018; FAO, 2011). The causes of FLW in low-income countries are mainly connected to technical, financial and managerial limitations in harvesting techniques as well as storage and cooling facilities (FAO, 2011). In industrialised regions (Europe, North America and industrialised Asia), losses in agricultural production are due to post-harvest grading (especially of fruit and vegetables) related to quality standards set by retailers (FAO, 2011; Johnson et al., 2018). Food may also be lost due to premature harvesting as poor farmers sometimes harvest crops too early due to desperate need for cash or food deficiency. Thus, food may get wasted if it is unsuitable for consumption (FAO, 2011). Losses in animal production are high in developing regions (e.g. FAO, 2011; Wesana et al., 2018). These losses are due, inter alia, to animal mortality, caused by

2 Food banks acquire donated food from different sources (e.g. retail stores, distributors, manufacturers, farms, consumers) that would otherwise be wasted. They then make donated food available through various community services (e.g. food pantries, soup kitchens, school feeding programmes, substance abuse clinics, hospices and other non-profit programmes) that provide food to those in need.

frequent diseases (e.g. parasites, pneumonia, digestive diseases) in livestock breeding. Similarly, dairy cow illnesses (e.g. mastitis) decrease milk production (FAO, 2011). Improvement in animal genetics and livestock production practices may increase efficiencies and reduce losses (e.g. Alexander et al., 2017). Likewise, breeding (e.g. cultivars more resistant to pest and diseases) can help in reducing pre- and post-harvest crop losses. Solutions to reduce FLW in production and harvesting also include good agricultural practices (GAPs)[3] as well as investments in mechanical harvesting (HLPE, 2014). It is also important to enhance the agronomic and managerial skills of farmers, which requires well-performing, effective and demand-driven extension and advisory services.

## 2.2 Transport, handling and storage

Losses due to poor transport and storage are more important in developing countries (FAO, 2011; HLPE, 2014). In fact, poor storage facilities and transport infrastructure cause post-harvest losses in developing countries. Fresh and perishable agro-food products (e.g. fruits, vegetables, meat, fish) can be spoilt in hot climates due to lack of infrastructure for storage, transportation and cooling (Rolle, 2006; Stuart, 2009). Also, insufficient or inadequate packaging can lead to FLW (Olsmats and Wallteg, 2009).

According to the International Institute of Refrigeration (2009), 23% of perishable agro-food products are lost in developing countries due to the lack of refrigeration use. Fruits and vegetables, plus roots and tubers have the highest wastage rates of any food. As for post-harvest losses, estimates range from 8% to 22% of cereals lost at farm-level and post-harvest due to poor storage to nearly 100% in some situations for horticultural produce (FAO, 2011; Parfitt et al., 2010). For perishable agro-foods, temperature management (e.g. cold chain) is a key element for FLW reduction (Badia-Melis et al., 2018; Mercier et al., 2018). Cold-chain management often depends on different interventions (e.g. pre-cooling, cold storage, refrigerated transport and refrigerated display during marketing) that involve actors all along the food chain (Albisu, 2014; HLPE, 2014). According to Gligor et al. (2018), key obstacles to the implementation of cold chain in Vietnam are

> deficient professional skills, lack of quality and safety-control measures, high concentration of intermediaries, poor infrastructure, lack of information systems, high cost of installation and operation, inadequate education and training at farmer level, deficiency of standardization, lack of government support for local businesses and social norms. (p. 942)

3 For example Integrated Crop Management (ICM), Integrated Pest Management (IPM) and so on.

FAO (2015a) points out that the causes of FLW in the Near East and North Africa (NENA) region are attributed to poor farming systems, deficient infrastructure and inadequate post-harvest practices (e.g. handling practices, drying techniques, cold storage). In several NENA countries, most of the farmers still use traditional methods in grains storage, with insects, rodents and birds being responsible for the bulk of cereal losses. Furthermore, the NENA region suffers from very low cold-chain capacity, especially due to its hot climate. The lack or unreliability of power supply is a key challenge to establishing the cold chain in the region. Poor maintenance and management practices are another major factor concerning the infrastructure in the region. Other types of infrastructure face similar major constraints.

Unsafe food – that does not comply with minimum food safety standards – is wasted as it is not fit for human consumption. A range of factors can lead to food being unsafe, such as unhygienic handling and storage conditions, inadequate temperature control, naturally occurring toxins, contaminated water as well as unsafe use of pesticides and veterinary drug residues (FAO, 2011).

Solutions to post-harvest losses range from investment in storage facilities to the adoption of innovative techniques and practices in transport, refrigeration, packaging and processing (HLPE, 2014). Logistical improvements and advances in cold-chain management are also enumerated among food wastage prevention solutions (e.g. European Commission, 2010; Porat et al., 2018). According to Foresight (2011), reducing post-harvest losses and waste can be achieved through deployment of existing technology and knowledge in storage and transport infrastructure as well as investment in new, appropriate technologies. Solutions to reduce FLW include investments in storage facilities and cold chain (HLPE, 2014). Governments should improve the road, energy and markets' infrastructure. Subsequently, private sector investments can improve transportation as well as cold-chain and storage facilities (Choudhury, 2006). Technological solutions such as wireless sensor networks (WSN), thermal imaging and radio frequency (RF) technologies (Badia-Melis et al., 2018), Internet-of-things (Badia-Melis et al., 2018; Pal and Kant, 2018) as well as biosensors (Neethirajan et al., 2018) can help to address cold-chain inefficiencies. Collective storage is also a solution for food losses reduction. However, its effectiveness depends on the local institutional context, such as existence of local cooperatives or producers' organisations (HLPE, 2014). Likewise, marketing cooperatives provide a central point for assembling produce from small farmers and preparing commodities for transportation to markets or other distribution channels. They should be able to reduce food losses by increasing the efficiency of these activities (Kader, 2005).

It is also necessary to develop capacity and knowledge of food chain operators to apply safe handling practices. Foods need to be produced, stored and handled in accordance with food safety standards. This requires the application of good agricultural and good hygienic practices by all food chain operators (FAO, 2011). Moreover, 'New technologies are required to reduce postharvest disease losses, without exposing consumers to hazardous chemical residues'[4] (Siddique et al., 2018:193).

Food wastage can also be minimised by modifying packaging and through other approaches that prolong the shelf life (WRAP, 2012). Using appropriate packaging may reduce food losses at almost every stage of the food chain (Olsmats and Wallteg, 2009). Moreover, packaging that carries information on how food should be conserved/stored also leads to food wastage reduction (HLPE, 2014).

## 2.3 Food processing

The lack of processing facilities causes food losses. In many developing countries, the food-processing industry does not have the capacity to preserve and process fresh farm produce due, among other reasons, to production seasonality and the high cost of processing facilities, especially when used only during a short-year period (FAO, 2011). Food is also lost during processing. Food-processing lines often carry out trimming to make sure that the end product is in the right size and shape. Trimmings are usually disposed of. Food is also lost during the processing line because of spoilage or some errors. In fact, end products with the wrong shape, appearance, weight or damaged packaging often end up being discarded, although taste or nutritional value are not affected (Stuart, 2009; Swedish Environmental Protection Agency, 2008).

Solutions to reduce FLW include good manufacturing practices (GMPs) and good hygienic practices (GHPs) (HLPE, 2014). Preservation processes such as pasteurisation, canning and sterilisation as well as packaging technologies contribute to increasing the shelf life of products, thereby reducing FLW (Langelaan et al., 2013). In addition to cold storage/chain, controlled atmosphere packaging and modified atmosphere packaging have been used successfully in decreasing metabolic activity of perishable agro-food products (e.g. fruit, vegetables) – that leads to ripening, over-ripening, senescence and loss of products – thus extending their shelf life (Falagán and Terry, 2018). Apart from 'hard' technological solutions, also 'soft' social and organisational ones are crucial in the fight against food wastage. Developing better linkages between processors and farmers (e.g. contract farming) can also help in reducing FLW. Furthermore, governments should create a better investment climate to stimulate the private sector to invest in the food-processing industry

4 These may include UV radiation, nanotechnology and plasma treatments as well as physical and biological pest control methods.

(FAO, 2011). Food-processing development requires appropriate processing technologies and infrastructure in a concerted food chain approach (HLPE, 2014).

## 2.4 Distribution, retail and marketing

At retail level, large quantities of food are wasted due to quality standards that over-emphasise appearance (e.g. cosmetic specifications). Some produce is rejected by supermarkets at farm due to rigorous standards concerning size, weight, appearance and shape of crops (de Hooge et al., 2018; Stuart, 2009). In fact, 'food retailers claim to address food waste while simultaneously setting standards that result in the large-scale rejection of edible food on cosmetic grounds' (Devin and Richards, 2018:199). Therefore, large portions of harvest never leave farms (Stuart, 2009).

Inadequate market systems in developing countries cause high food losses. There are too few wholesale markets, supermarket and retail facilities providing suitable sale conditions for agro-food products. Markets (both wholesale and retail ones) in developing countries are often overcrowded, small and unsanitary (Kader, 2005). For instance, markets in NENA region often lack cooling equipment and adequate facilities for loading, unloading, ripening, consumer packaging and temporary storage (FAO, 2015b).

Large quantities and a wide range of products on display in supermarkets lead to food wastage. Consumers, especially in industrialised countries, expect a variety of food types and brands to be available in stores, which increases the likelihood of some of these products reaching their 'sell-by' date before being sold, and thereby discarded. Moreover, food products that are close to their expiry are often ignored by consumers (Swedish Environmental Protection Agency, 2008).

Selling farm crops closer to consumers (e.g. farmers' markets and farm shops), without passing the strict standards set up by supermarkets, would reduce the amount of rejected agricultural products (Stuart, 2009). Moreover, charity organisations could collect and use discarded 'substandard' products, as long as they are still safe and of good nutritional value (Swedish Environmental Protection Agency, 2008). Substandard agro-food products can be also sold in stores (de Hooge et al., 2018; Stuart, 2009). Surveys show that consumers are willing to buy heterogeneous produce (in terms of size, weight, shape, appearance) as long as the taste is not affected (Stuart, 2009). Aschemann-Witzel (2018a) suggests that

> Consumer acceptance of expiration date based pricing of suboptimal food can be increased through furthering consumer familiarity with the practice, improving perceived quality and providing tips to ensure consumers are confident to be able to use the entire food at home. (p. 119)

## 2.5 Consumption

Every year, consumers in rich countries waste almost as much food (222 million tonnes) as the entire net food production of sub-Saharan Africa (230 million tonnes) (HLPE, 2014). The causes of food wastage in medium/high-income countries mainly relate to consumer behaviour (e.g. FAO, 2011; Gaiani et al., 2018; Porat et al., 2018). In fact, inadequate purchase planning (e.g. lack of use of shopping list) and confusing 'best-before' expiry dates, in combination with the careless attitude of consumers that can afford to waste food, cause large amounts of food wastage in medium/high-income countries (FAO, 2011). In fact, as TEEB (2018) puts it, 'disposing is cheaper than reusing food in industrialized food chains and food wastage proportions continue to grow in parallel to increasing production and consumption' (p. 35). Careless consumer attitude is exacerbated by the abundance of food in developed countries; the amount of available food per person has increased steadily during the last decades in both the United States and Europe. Many restaurants serve nowadays buffets at fixed prices (e.g. all-you-can-eat buffets) and this may encourage clients to fill their plates with more food than they would actually eat. Principato et al. (2018:130) argue that '... the attitude of restaurant managers as well as types of menus served and restaurant size play significant roles' in determining the amount of food wasted in restaurants. Likewise, large portion sizes – that might be due to inappropriate public tenders – can lead to food wastage in school canteens (Balzaretti et al., 2018). Moreover, retail stores and supermarkets promote 'buy one, get one free' bargains and offer large packages. Also food industry tends to produce oversized ready-to-eat meals (Stuart, 2009).

Overconsumption[5] is also considered as a kind of food wastage (Alexander et al., 2017; Blair and Sobal, 2006; Schmidt and Matthies, 2018; Searchinger et al., 2013). In fact, reducing the consumption of excess calories would reduce the 2050 calorie gap by 6% (Searchinger et al., 2013). Alexander et al. (2017) found over-eating to be 'at least as large a contributor to food system losses as consumer food waste' (p. 190). Some scholars also consider consumption of resource-intensive food items instead of equally nutritious, more efficient alternatives as food wastage (e.g. Shepon et al., 2018).[6] From this perspective, Alexander et al. (2017) consider that 'The highest rates of loss are associated with livestock production'[7] (p. 190), which – combined with the health and climate change benefits of dietary change (e.g. Springmann et al., 2016) – makes the case for a reduction of the consumption of animal-based foods.

5 Additional food intake over that required/recommended for good human nutrition.

6 Shepon et al. (2018) define 'opportunity food losses as the food loss associated with consuming resource-intensive animal-based items instead of plant-based alternatives which are nutritionally comparable, e.g., in terms of protein content' (p. 3804).

7 Alexander et al. (2017) consider losses in livestock production as 'Losses and inefficiencies in the conversion of feed and grass into animal products' (p. 191).

Consumer food waste reduction implies implementation of technical and behaviour-driven solutions in food services (e.g. canteens, catering facilities, restaurants etc.), hospitality sector (e.g. hotels etc.) and households (HLPE, 2014). In fact, food wastage in rich countries can be reduced by raising awareness among retailers, consumers and food industries (Aschemann-Witzel, 2018b; FAO, 2011). Solutions to reduce FLW should be tailored taking into account the social and cultural context of each area. For instance, significant waste takes place in the NENA region also during social events (e.g. wedding ceremonies, family gatherings) and religious occasions such as the fasting month of Ramadan (FAO, 2015b). Education on food waste in schools and educational institutions (e.g. Pinto et al., 2018) as well as political initiatives are potential starting points to change people's attitude (FAO, 2011). Von Kameke and Fischer (2018) suggest that 'nudging' - that is more likely to lead to concrete, perceptible behaviour change - can help in preventing household food waste and argue that it is more effective than awareness campaigns. Food sharing/redistribution and food banks can also help to reduce food wastage at consumption level (e.g. Bucatariu, 2016; Facchini et al., 2018).

The Bio Intelligence Service (European Commission, 2010) identified a wide range of food waste prevention solutions: informational tools, training programmes, regulatory measures, food redistribution programmes and awareness campaigns. Quested et al. (2013) detail measures that can be implemented by consumers to reduce food waste: understanding better the difference between 'best-before' and 'use-by' dates; avoiding buying more than is needed by better purchases planning; estimating better the portions that need to be prepared when cooking; avoiding advance or impulsive purchasing of food that is not immediately needed; following better storage practices at home; possessing better knowledge of recipes to use leftovers.

## 3 Reducing food wastage along the food supply chain: the need for a systems approach

Food wastage reduction could have multiple sustainability benefits, including reduced environmental impacts and improved nutritional status (especially for low-income groups). Nevertheless, realising such benefits require a systemic approach based on a good understanding of the multiple underlying factors (including policies and regulatory framework) that affect how agro-food systems are structured, how they operate and, most importantly, why they generate such amounts of surplus/wasted food (Bengtsson et al., 2018). Such a 'systems approach' should allow comprehending the functioning of the food system, from production to consumption, in order to identify appropriate leverage points to bring about FLW reduction. In fact, 'Food wastage is intrinsically linked to the way food systems function technically, economically and culturally' (TEEB, 2018:35).

Efficient solutions exist to reduce FLW along the whole food chain. Actions to reduce food wastage should not be directed towards isolated, single parts of the food chain, since what is done in one part of the chain has effects in other ones (FAO, 2011). For instance, Devin and Richards (2018) argue that food retailers '... may be considered to be legitimately engaging in socially responsible behaviors to lower food waste, yet implement practices that ultimately contribute to higher levels of food waste elsewhere in the supply chain' (p. 199). HLPE (2014) distinguishes three levels of FLW causes: micro, meso[8] and macro.[9] Meso- and macro-levels are important as quite often, causes of FLW of a physical, technical or behavioural nature at micro-level (e.g. household) are induced by a broader institutional, economic and/or social context. Categories of solutions to reduce food wastage include good practices, investments, behavioural change, governance of the food chain and policy coordination (HLPE, 2014). The supply chains in developing countries need to be strengthened by, among others, encouraging small-scale farmers to better organise their production and marketing. Investments in transportation, infrastructure, packaging and processing are also needed. Both the private and public sectors have an important role to play in achieving this (FAO, 2011). HLPE (2014) identified a host of collective, concerted actions to reduce food wastage along the food chain (Table 1).

The causes of FLW in medium/high-income countries mainly relate to a lack of coordination between different actors in the food supply chain (FAO, 2011). In fact, food gets lost because production exceeds demand. Farmers try to anticipate pest attacks and unpredictable, bad weather while ensuring delivery of quantities of produce agreed with retailers or processors. Therefore, they sometimes end up having surplus crop that is often discarded, unless it is sold to processors or used as animal feed, which is not always financially profitable (FAO, 2011). Better cooperation and communication between farmers, especially smallholders, can decrease overproduction risk as surplus produce from one farm can be used to solve shortage on another one (Stuart, 2009). For that, small resource-poor farmers need to be organised in groups (FAO, 2011). Producers' organisations (e.g. associations, cooperatives) as well as their federations can help to reduce losses through management and organisational innovations, for example by supporting collective planning of production, sorting and grading activities as well as logistics (Despoudi et al., 2018; FAO, 2012b; FAO, IFAD and WFP, 2012). Furthermore, farmer-buyer sale agreements (e.g. contract farming) may contribute to reducing

8 These include lack of support for investments and good practices, lack of infrastructure, lack of integrated food chain management approaches and confusion around food date labelling (e.g. best before, use by).
9 These refer to the impact of policies, laws and regulations as well as systemic causes (e.g. financial, managerial and technical limitations; lack of infrastructure; lack of coordination among supply chain actors; consumer behaviour).

**Table 1** Examples of concerted and collective solutions to reduce food losses and waste by different stages in the food chain

| Stage | Actions |
|---|---|
| Pre-harvest/ production stage | Strengthening producer organisations/associations in good agricultural practices, good harvest practices and so on.<br>Improving availability and quality of agricultural extension services, especially for small-scale farmers.<br>Training farmers on good harvest practices (e.g. produce maturity indices, harvesting time).<br>Fostering horizontal integration of farmers (e.g. farmer organisations/ cooperatives) to better plan supply and reduce overproduction risk.<br>Improving vertical integration within the value chain (e.g. contract farming) to improve efficiency.<br>Making easier and financially viable the use of unmarketable agro-foods in production areas (e.g. cottage processing industries, donation). |
| Handling and storage | Improving farmers' access to low-cost storage and handling technologies (e.g. storage bags, evaporative coolers, metallic silos).<br>Building capacity of primary producers and their organisations in good storage practices and good hygienic practices.<br>Training food supply chain operators (e.g. growers, traders, transporters) on good post-harvest handling technologies and good storage practices (e.g. ethylene and microbial management).<br>Ensuring safety and pest control protocols are followed along the whole value chain.<br>Promoting joint public-private investments to improve storage facilities, especially in developing countries.<br>Supporting joint/group storage facilities for small-scale farmers. |
| Processing and packaging | Enhancing skills of food chain operators in relation to good manufacturing practices.<br>Promoting public-private partnerships in food processing.<br>Creating an enabling environment for investments in agro-food processing by the private sector.<br>Strengthening and/or developing linkages between farmers and processors.<br>Enhancing supply chain management to improve the supply/demand balance and overcome the seasonality of processing facilities operation.<br>Ensuring the adherence of processors to standards set for processed foods to avoid producing substandard products. |
| Transport, distribution and marketing | Facilitating producers' access to markets.<br>Developing markets for substandard/suboptimal agro-food products.<br>Promoting commodity groups (cooperatives, associations, organisations) to improve efficiency and market access.<br>Providing guidance on good food storage and preparation to consumers.<br>Clarifying food date labelling (e.g. sell-by date) to avoid misunderstanding by consumers.<br>Facilitating donation of unsold food products (e.g. food banking). |
| Consumption | Educating consumers on better planning of their buying (e.g. use of shopping list) and on buying only according to planned meals.<br>Conducting awareness-raising campaigns and consumer education on good storage practices, reuse of leftovers, proper interpretation of labelling dates.<br>Fighting against the marketing practices that 'devalorize' food (e.g. buy one, get one free). |

*Source*: adapted from HLPE (2014).

quantities of farm crops being lost (FAO, 2011). FLW reduction also implies new ways of governance and organisation of agro-food supply chains. This is the case for closed-loop supply chain models (World Economic Forum, 2010, 2014) whose development is meant to coordinate actions of all actors towards FLW reduction. In such chain models, losses or waste of all forms are fed back into the supply chain (HLPE, 2014).

Also Lomax and Bortoletti (2018) highlight the need for a 'food systems approach to decision making' that implies the design and/or application of integrated and adaptive policy interventions, which are informed by a holistic diagnosis of constraints to sustainable food system development and based on an enhanced cooperation among food systems' actors. Likewise, HLPE (2014) points out that solutions to reduce FLW imply food chain actors' coordination as well as policy coordination. Janousek et al. (2018) put that 'To fully address food waste, actions oriented toward minimizing and sustainably managing food waste must be undertaken in a collaborative manner across all stages of the food supply chain' (p. 933). Lomax and Bortoletti (2018) add that such a 'food systems approach' needs to consider and adopt five interlinked principles: focus on food systems outcomes, include food consumption as a driver, facilitate platforms of collaboration among food systems actors, address emerging trends and challenges (e.g. population growth, urbanisation) and promote integration between food and agriculture policies and other agendas (e.g. health, environment). The integration of FLW concerns in policies can take two different, but complementary, ways: (1) integrating FLW concerns in all food-related policies; (2) devising a specific, transversal FLW reduction policy (HLPE, 2014).

Alamar et al. (2018) argue that there is 'a need for a paradigm shift of current funding strategies and research programmes that will encourage the development, implementation and translation of collective biological, engineering and management solutions to better preserve and utilise food', thus proposing to encourage inter-supply chain knowledge exchange, research and capacity building to reduce FLW (Alamar et al., 2018). There is also a need for a better integration of scientific disciplines and research strands. In fact, as Schanes et al. (2018) point out, food waste is '... a complex and multi-faceted issue that cannot be attributed to single variables; this also calls for a stronger integration of different disciplinary perspectives' (p. 978).

Solutions to reduce FLW at one stage of the food chain often involve actors upstream or downstream in the chain. Thus, the implementation of these solutions often requires coordination between various actors including farmers and their organisations, public organisations, processors, retailers, research institutes, civil society organisations and consumers and their organisations. The feasibility, effectiveness, efficiency and sustainability of actions for the reduction of food wastage in the medium/long term depend, therefore, on the level of coordination among food chain actors, which, in turn, is affected by the institutional environment and political context in each country/area.

## 4 Future trends and conclusion

Curbing the amount of FLW is a tangible starting point for moving towards sustainable food consumption and production, which is crucial to achieve long-term food security and sustainability. FLW represents one of the most critical environmental, social, economic and ethical challenges faced by humanity. Instead of current emphasis put on intensification for increasing food production to meet the growing food demand, more resource should be channelled towards enabling innovations and technologies to reduce FLW. Some important constraints have so far hindered effective actions in reducing FLW. These include, among others, lack of awareness among food supply chain actors; non-existent or inappropriate policies, regulations and incentives; institutional and coordination gaps; and insufficient investment (e.g. processing, cold-chain infrastructure, transportation). To foster commitment for FLW reduction, several crosscutting actions and strategies must be developed and implemented. FLW can be reduced effectively only with a systemic, integrated and holistic food chain approach that takes into account the multidimensional relations between the different food chain stages and actors. Moreover, food wastage reduction is a social and collective imperative that requires the commitment of the whole society, from government agencies to the private sector, civil society (e.g. NGOs), producer cooperatives/organisations, food industry and consumers.

Reduction of FLW through improvement of the food chain – thus fostering more efficient and smarter food production systems and consumption patterns – can contribute to achieving food security for all and improving the sustainability of the global food system with positive environmental, economic and social implications. This chapter makes the case for adopting a 'systems approach' in order to realise the full benefits of FLW reduction in terms of food security and sustainability. It also stresses that the improvement of the management and governance of the whole food chain is crucial to address the underlying causes of food wastage. This can be achieved through the deployment of adequate innovations (both technical and soft organisational/social ones) along the food chain. The combination of economic/fiscal, informational/communication, regulatory, behavioural and technological instruments is crucial to bring about an effective reduction or prevention of FLW.

Future research on FLW should address the causes, extent and implications (economic, social, environmental) of FLW along the food chain as well as the effectiveness of reduction strategies in different contexts and food environments. It should also pay more attention to the multifaceted relations between FLW and food and nutrition security especially in developing countries.

## 5 Where to look for further information

FAO - SAVE FOOD: global initiative on food loss and waste reduction: http://www.fao.org/save-food/en

FAO - Technical platform on the measurement and reduction of food loss and waste: http://www.fao.org/platform-food-loss-waste/en

UN Environment - Food and food waste: https://www.unenvironment.org/explore-topics/resource-efficiency/what-we-do/sustainable-lifestyles/food-and-food-waste

European Commission - EU Platform on food losses and food waste: https://ec.europa.eu/food/safety/food_waste/eu_actions/eu-platform_en

World Resources Institute (WRI) - Food waste: https://www.wri.org/tags/food-waste

OECD - Food waste: https://stats.oecd.org/Index.aspx?DataSetCode=FOOD_WASTE

## 6 References

Alamar, M. D. C., Falagán, N., Aktas, E. and Terry, L. A. 2018. Minimising food waste: a call for multidisciplinary research. *Journal of the Science of Food and Agriculture* 98(1), 8–11. doi:10.1002/jsfa.8708.

Albisu, L. M. 2014. Food losses and waste in the Spanish agro-food chain. CIHEAM Watch Letter No. 30, September 2014. Available at: www.ciheam.org/images/CIHEAM/PDFs/Publications/LV/wl30official.pdf.

Alexander, P., Brown, C., Arneth, A., Finnigan, J., Moran, D. and Rounsevell, M. D. A. 2017. Losses, inefficiencies and waste in the global food system. *Agricultural Systems* 153, 190–200. doi:10.1016/j.agsy.2017.01.014.

Aragie, E., Balié, J. and MoralesOpazo, C. 2018. Does reducing food losses and wastes in sub-Saharan Africa make economic sense? *Waste Management and Research: the Journal of the International Solid Wastes and Public Cleansing Association, ISWA* 36(6), 483–94. doi:10.1177/0734242X18770247.

Aschemann-Witzel, J. 2018a. Consumer perception and preference for suboptimal food under the emerging practice of expiration date based pricing in supermarkets. *Food Quality and Preference* 63, 119–28. doi:10.1016/j.foodqual.2017.08.007.

Aschemann-Witzel, J. 2018b. Helping you to waste less? Consumer acceptance of food marketing offers targeted to food-related lifestyle segments of consumers. *Journal of Food Products Marketing* 24(5), 522–38. doi:10.1080/10454446.2018.1472693.

Badia-Melis, R., McCarthy, U., Ruiz-Garcia, L., Garcia-Hierro, J. and Robla Villalba, J. I. 2018. New trends in cold chain monitoring applications – a review. *Food Control* 86, 170–82. doi:10.1016/j.foodcont.2017.11.022.

Balzaretti, C. M., Ventura, V., Ratti, S., Ferrazzi, G., Spallina, A., Carruba, M. O. and Castrica, M. 2018. Improving the overall sustainability of the school meal chain: the role of portion sizes. *Eating and Weight Disorders: EWD*. doi:10.1007/s40519-018-0524-z.

Bengtsson, M., Alfredsson, E., Cohen, M., Lorek, S. and Schroeder, P. 2018. Transforming systems of consumption and production for achieving the sustainable development goals: moving beyond efficiency. *Sustainability Science* 13(6), 1533–47. doi:10.1007/s11625-018-0582-1.

Blair, D. and Sobal, J. 2006. Luxus consumption: wasting food resources through overeating. *Agriculture and Human Values* 23(1), 63–74. doi:10.1007/s10460-004-5869-4.

Bucatariu, C. 2016. Food recovery and redistribution: identification and monitoring of (i) societal needs, (ii) institutional frameworks, and (iii) options for operational implementation. In: Meybeck, A. and Redfern, S. (Eds), *Knowledge and Information for Sustainable Food Systems. Proceedings of a Workshop of the FAO/UNEP Programme on Sustainable Food Systems*, 10–11 September 2014, Rome, Italy. FAO, Rome, pp. 133–44. Available at: www.fao.org/3/a-i5373e.pdf.

Choudhury, M. L. 2006. Recent developments in reducing postharvest losses in the Asia-Pacific region. In: Rolle, R. S. (Ed.), *Postharvest Management of Fruit and Vegetables in the Asia-Pacific Region*, Tokyo. Asian Productivity Organization, pp. 15–22. Available at: https://www.apo-tokyo.org/00e-books/AG-18_PostHarvest/AG-18_PostHarvest.pdf.

Corrado, S. and Sala, S. 2018. Food waste accounting along global and European food supply chains: state of the art and outlook. *Waste Management* 79, 120–31. doi:10.1016/j.wasman.2018.07.032.

de Hooge, I. E., van Dulm, E. and van Trijp, H. C. M. 2018. Cosmetic specifications in the food waste issue: supply chain considerations and practices concerning suboptimal food products. *Journal of Cleaner Production* 183, 698–709. doi:10.1016/j.jclepro.2018.02.132.

Despoudi, S., Papaioannou, G., Saridakis, G. and Dani, S. 2018. Does collaboration pay in agricultural supply chain? An empirical approach. *International Journal of Production Research* 56(13), 4396–417. doi:10.1080/00207543.2018.1440654.

Devin, B. and Richards, C. 2018. Food waste, power, and corporate social responsibility in the Australian food supply chain. *Journal of Business Ethics* 150(1), 199–210. doi:10.1007/s10551-016-3181-z.

Di Terlizzi, B., Van Otterdijk, R., Dragotta, A., Pink, P. and El Bilali, H. 2016. Innovation for the reduction of food losses and waste. In: International Centre for Advanced Mediterranean Agronomic Studies (CIHEAM) and Food and Agriculture Organisation of the United Nations (FAO) (Eds), *Mediterra 2016 'Zero Waste in the Mediterranean: Natural Resources, Food and Knowledge'*. Presses des SciencesPo, Paris, pp. 281–301. Available at: https://www.ciheam.org/uploads/attachments/333/Mediterra2016_EN_BAT__1_.pdf.

European Commission. 2010. Preparatory study on food waste across EU 27. Brussels. doi:10.2779/85947.

Facchini, E., Iacovidou, E., Gronow, J. and Voulvoulis, N. 2018. Food flows in the United Kingdom: the potential of surplus food redistribution to reduce waste. *Journal of the Air and Waste Management Association* 68(9), 887–99. doi:10.1080/10962247.2017.1405854.

Falagán, N. and Terry, L. A. 2018. Recent advances in controlled and modified atmosphere of fresh produce. *Johnson Matthey Technology Review* 62(1), 107–17. doi:10.1595/205651318X696684.

FAO. 2011. Global food losses and food waste: extent, causes and prevention. FAO, Rome. Available at: www.fao.org/docrep/014/mb060e/mb060e.pdf.

FAO. 2012a. Greening the economy with agriculture. FAO, Rome. Available at: www.fao.org/docrep/015/i2745e/i2745e00.pdf.

FAO. 2012b. Smallholder business models for agribusiness-led development: good practice and policy guidance. FAO, Rome. Available at: www.fao.org/docrep/015/md923e/md923e00.pdf.

FAO. 2012c. Towards the future we want: end hunger and make the transition to sustainable agricultural and food systems. FAO, Rome. Available at: www.fao.org/docrep/015/an894e/an894e00.pdf.

FAO. 2013. Food wastage footprint. Impacts on natural resources. FAO, Rome. Available at: www.fao.org/docrep/018/i3347e/i3347e.pdf.

FAO. 2015a. Regional overview of food insecurity – Near East and North Africa: strengthening regional collaboration to build resilience for food security and nutrition. FAO, Cairo.

FAO. 2015b. Regional strategic framework – reducing food losses and waste in the Near East & North Africa region. FAO, Cairo.

FAO. 2017. *The Future of Food and Agriculture: Trends and Challenges*. FAO, Rome: ISBN: 978-92-5-109551-5.

FAO, IFAD and WFP. 2012. The role of producer organizations in reducing food loss and waste – International Year of Cooperatives Issue Brief Series. Available at: www.fao.org/docrep/016/ap409e/ap409e.pdf.

FAO, IFAD, UNICEF, WFP and WHO. 2017. State of food insecurity and nutrition in the world – building resilience for peace and food security. FAO, Rome. Available at: www.fao.org/3/a-I7695e.pdf.

Foresight. 2011. The future of food and farming: challenges and choices for global sustainability. Final Project Report. The Government Office for Science, London, UK.

Gaiani, S., Caldeira, S., Adorno, V., Segrè, A. and Vittuari, M. 2018. Food wasters: profiling consumers' attitude to waste food in Italy. *Waste Management* 72, 17–24. doi:10.1016/j.wasman.2017.11.012.

Gligor, D., Tan, A. and Nguyen, T. N. T. 2018. The obstacles to cold chain implementation in developing countries: insights from Vietnam. *The International Journal of Logistics Management* 29(3), 942–58. doi:10.1108/IJLM-02-2017-0026.

HLPE. 2011. Price volatility and food security. A report by the High Level Panel of Experts on Food Security and Nutrition (HLPE) of the Committee on World Food Security. HLPE, Rome. Available at: www.fao.org/3/a-av038e.pdf.

HLPE. 2014. Food losses and waste in the context of sustainable food systems. A report by the High Level Panel of Experts on Food Security and Nutrition (HLPE) of the Committee on World Food Security. HLPE, Rome. Available at: http://www.fao.org/3/a-i3901e.pdf.

Ingram, J. S. I. 2011. *From Food Production to Food Security: Developing Interdisciplinary, Regional-Level Research*. Wageningen University, Wageningen, The Netherlands.

International Institute of Refrigeration. 2009. *The Role of Refrigeration in Worldwide Nutrition*. International Institute of Refrigeration, Paris.

Janousek, A., Markey, S. and Roseland, M. 2018. 'We see a real opportunity around food waste': exploring the relationship between on-farm food waste and farm characteristics. *Agroecology and Sustainable Food Systems* 42(8), 933–60. doi:10.1080/21683565.2018.1468381.

Johnson, L. K., Dunning, R. D., Bloom, J. D., Gunter, C. C., Boyette, M. D. and Creamer, N. G. 2018. Estimating on-farm food loss at the field level: a methodology and applied case study on a North Carolina farm. *Resources, Conservation and Recycling* 137, 243–50. doi:10.1016/j.resconrec.2018.05.017.

Kader, A. A. 2005. Increasing food availability by reducing postharvest losses of fresh produce. *Acta Horticulturae* 682(682), 2169–76. doi:10.17660/ActaHortic.2005.682.296.

Kummu, M., de Moel, H., Porkka, M., Siebert, S., Varis, O. and Ward, P. J. 2012. Lost food, wasted resources: global food supply chain losses and their impacts on freshwater, cropland, and fertiliser use. *The Science of the Total Environment* 438, 477–89. doi:10.1016/j.scitotenv.2012.08.092.

Langelaan, H. C., Pereira da Silva, F., Thoden van Velzen, U., Broeze, J., Matser, A. M., Vollebregt, M. and Schroën, K. 2013. Technology options for feeding 10 billion people – options for sustainable food processing. State of the art report, Science and Technology Options Assessment. European Union, Brussels. Available at: www.europarl.europa.eu/RegData/etudes/etudes/join/2013/513533/IPOLJOIN_ET(2013)513533_EN.pdf.

Lipinski, B., Hanson, C., Lomax, J., et al. 2016. Reducing food loss and waste. Installment 2 of 'Creating a Sustainable Food Future'. Working Paper. World Resources Institute, Washington DC. doi:10.2499/9780896295827_03.

Lomax, J. and Bortoletti, M. 2018. *Transformative Sustainable Food Systems Framework*. UN Environment, Paris.

Mercier, S., Mondor, M., Villeneuve, S. and Marcos, B. 2018. The Canadian food cold chain: A legislative, scientific, and prospective overview. *International Journal of Refrigeration* 88, 637–45. doi:10.1016/j.ijrefrig.2018.01.006.

Neethirajan, S., Ragavan, V., Weng, X. and Chand, R. 2018. Biosensors for sustainable food engineering: challenges and perspectives. *Biosensors* 8(1), 23. doi:10.3390/bios8010023.

Olsmats, C. and Wallteg, B. 2009. Packaging is the answer to world hunger. Available at: www.worldpackaging.org/i4a/pages/index.cfm?pageid=1#&panel1-1.

Pal, A. and Kant, K. 2018. IoT-based sensing and communications infrastructure for the fresh food supply chain. *Computer* 51(2), 76–80. doi:10.1109/MC.2018.1451665.

Parfitt, J., Barthel, M. and Macnaughton, S. 2010. Food waste within food supply chains: quantification and potential for change to 2050. *Philosophical Transactions of the Royal Society of London. Series B, Biological Sciences* 365(1554), 3065–81. doi:10.1098/rstb.2010.0126.

Pinto, R. S., Pinto, R. M. D. S., Melo, F. F. S., Campos, S. S. and Cordovil, C. M. 2018. A simple awareness campaign to promote food waste reduction in a university canteen. *Waste Management* 76, 28–38. doi:10.1016/j.wasman.2018.02.044.

Porat, R., Lichter, A., Terry, L. A., Harker, R. and Buzby, J. 2018. Postharvest losses of fruit and vegetables during retail and in consumers' homes: quantifications, causes, and means of prevention. *Postharvest Biology and Technology* 139, 135–49. doi:10.1016/j.postharvbio.2017.11.019.

Principato, L., Pratesi, C. A. and Secondi, L. 2018. Towards zero waste: an exploratory study on restaurant managers. *International Journal of Hospitality Management* 74, 130–7. doi:10.1016/j.ijhm.2018.02.022.

Quested, T. E., Marsh, E., Stunell, D. and Parry, A. D. 2013. Spaghetti soup: the complex world of food waste behaviours. *Resources, Conservation and Recycling* 79, 43–51. doi:10.1016/j.resconrec.2013.04.011.

Rolle, R. S. 2006. Improving postharvest management and marketing in the Asia-Pacific region: issues and challenges. In: Rolle, R. S. (Ed.), *Postharvest Management of Fruit and Vegetables in the Asia-Pacific Region*, Tokyo, 2006. Asian Productivity

Organization, pp. 23–31. Available at: https://www.apo-tokyo.org/00e-books/AG-18_PostHarvest/AG-18_PostHarvest.pdf.
Schanes, K., Dobernig, K. and Gözet, B. 2018. Food waste matters – a systematic review of household food waste practices and their policy implications. *Journal of Cleaner Production* 182, 978–91. doi:10.1016/j.jclepro.2018.02.030.
Scherhaufer, S., Moates, G., Hartikainen, H., Waldron, K. and Obersteiner, G. 2018. Environmental impacts of food waste in Europe. *Waste Management* 77, 98–113. doi:10.1016/j.wasman.2018.04.038.
Schmidt, K. and Matthies, E. 2018. Where to start fighting the food waste problem? Identifying most promising entry points for intervention programs to reduce household food waste and overconsumption of food. *Resources, Conservation and Recycling* 139, 1–14. doi:10.1016/j.resconrec.2018.07.023.
Searchinger, T., Hanson, C., Ranganathan, J., Lipinski, B., Waite, R., Winterbottom, R., Dinshaw, A. and Heimlich, R. 2013. Creating a sustainable food future: A menu of solutions to sustainably feed more than 9 billion people by 2050. World resources report 2013–14: interim findings. World Resources Institute (WRI), Washington DC.
Shepon, A., Eshel, G., Noor, E. and Milo, R. 2018. The opportunity cost of animal based diets exceeds all food losses. *Proceedings of the National Academy of Sciences of the United States of America* 115(15), 3804–9. doi:10.1073/pnas.1713820115.
Siddique, S. S., Hardy, G. E. S. J. and Bayliss, K. L. 2018. Advanced technologies for controlling postharvest diseases of fruit. *Acta Horticulturae* 1194(1194), 193–200. doi:10.17660/ActaHortic.2018.1194.29.
Smil, V. 2004. Improving efficiency and reducing waste in our food system. *Environmental Sciences* 1(1), 17–26. doi:10.1076/evms.1.1.17.23766.
Springmann, M., Godfray, H. C. J., Rayner, M. and Scarborough, P. 2016. Analysis and valuation of the health and climate change cobenefits of dietary change. *Proceedings of the National Academy of Sciences of the United States of America* 113(15), 4146–51. doi:10.1073/pnas.1523119113.
Stuart, T. 2009. *Waste: Uncovering the Global Food Scandal*. W.W. Norton Co, London, UK.
Swedish Environmental Protection Agency. 2008. *Svinn I livsmedelskedjan – möjligheter till minskade mängder* [*Wastage in the Food Chain – Reduction Strategies*]. Swedish Environmental Protection Agency, Bromma, Sweden.
TEEB. 2018. Measuring what matters in agriculture and food systems: a synthesis of the results and recommendations of TEEB for Agriculture and Food's Scientific and Economic Foundations report. TEEB (The Economics of Ecosystems and Biodiversity) and UN Environment, Geneva. Available at: http://teebweb.org/agrifood/measuring-what-matters-in-agriculture-and-food-systems.
United Nations. 2015. Transforming our world: the 2030 Agenda for Sustainable Development. Resolution adopted by the General Assembly on 25 September 2015. United Nations, New York, NY. Available at: http://www.un.org/ga/search/view:doc.asp?symbol=A/RES/70/1&Lang=E.
Usubiaga, A., Butnar, I. and Schepelmann, P. 2018. Wasting food, wasting resources: potential environmental savings through food waste reductions. *Journal of Industrial Ecology* 22(3), 574–84. doi:10.1111/jiec.12695.
von Kameke, C. and Fischer, D. 2018. Preventing household food waste via nudging: an exploration of consumer perceptions. *Journal of Cleaner Production* 184, 32–40. doi:10.1016/j.jclepro.2018.02.131.

Wesana, J., De Steur, H., Dora, M. K., Mutenyo, E., Muyama, L. and Gellynck, X. 2018. Towards nutrition sensitive agriculture. Actor readiness to reduce food and nutrient losses or wastes along the dairy value chain in Uganda. *Journal of Cleaner Production* 182, 46–56. doi:10.1016/j.jclepro.2018.02.021.

World Economic Forum. 2010. Driving sustainable consumption. Closed loop systems – overview briefing. World Economic Forum, Cologny, Switzerland. Available at: www.weforum.org/pdf/sustainableconsumption/DSC Overview Briefing - Closed Loop Systems.pdf.

World Economic Forum. 2014. Towards the circular economy: accelerating the scale-up across global supply chains. World Economic Forum, Cologny, Switzerland. Available at: www3.weforum.org/docs/WEF_ENV_TowardsCircularEconomy_Report_2014.pdf.

WRAP. 2012. Courtauld Commitment 2 Voluntary Agreement 2010–2012 – signatory case studies and quotes. Waste Resources Action Programme (WRAP), Banbury, UK.

Wunderlich, S. M. and Martinez, N. M. 2018. Conserving natural resources through food loss reduction: production and consumption stages of the food supply chain. *International Soil and Water Conservation Research* 6(4), 331–9. doi:10.1016/j.iswcr.2018.06.002.

# Chapter 5

# Developing decision support systems for improving data management in agricultural supply chains

*Gerhard Schiefer, University of Bonn, Germany*


1 Introduction
2 Decisions in supporting data management
3 Decision tools
4 Principal case studies
5 Conclusion and future trends
6 References


## 1 Introduction

Food production involves one of the most complex organizational models of enterprise cooperation. The core is constituted by the actors involved in the flow of food products reaching from production agriculture to various levels of food processing, trade and retail, which, in addition, reach from rural agricultural regions to urban environments.

The actors involved in the production flow are summarized as members of an agricultural or food supply value chain. A food supply value chain (or shorter: a food value chain or food chain, a term used in the remainder of the chapter) is a theoretical construct for demonstrating the flow of products from agriculture to retail and consumers. However, a food value chain with a fixed group of enterprises with a continuous trade relationship is more the exception than the rule. The usual food chain view is comprised of a network of enterprises with dynamically changing trade relationships reaching from agriculture to consumers, resulting in a dynamically changing organization of food supply chains and their participating enterprises (Fig. 1).

However, whatever the flow of products, coordination is needed for assuring that production and distribution processes along the chain deliver

http://dx.doi.org/10.19103/AS.2020.0069.20

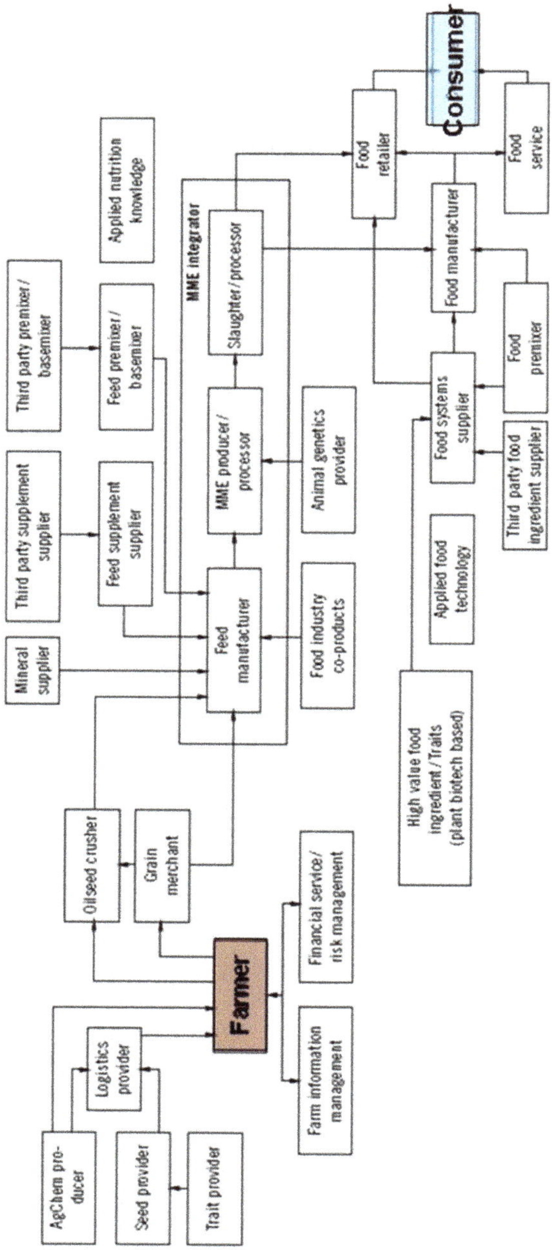


**Figure 1** Model of a food value chain. Source: author.

certain product qualities at consumers' end or interact with the environment for environmental protection according to society's interests. These 'classical' coordination needs have been discussed among stakeholders or interested groups in business, policy and research since a long time ago.

With the increasing mutual dependency of enterprises within the food chain especially in quality production and the emerging discussions on sustainability, the flow of data along the chain and their coordination are receiving increasing attention not the least because of increasing requirements for transparency along the food chain and toward consumers. Data coordination has proved to be quite complex. The point of data use might differ from the point of data collection. A critical point concerns agriculture as many of the data needed at later stages of the chain depend on issues where agriculture could provide information. This is the basis for coordination needs, which do not just have to agree on data collection and data use but also on agreements regarding the allocation of costs and benefits.

The flow of data is closely connected with the sequence of decisions that are being made along the food chain. Decisions require data, and the functioning of the sequence of decisions depends on the availability of data where needed. Decisions at one stage of the food chain might provide data for subsequent stages and influence activities and decisions at other stages. Coordination needs have to assure that the data for decisions are available where and when needed and that decisions do not contradict each other but contribute to the realization of joint objectives.

This constellation characterizes the difficulties in the management of data across the food value chain or in arriving at a coordination of decisions that serve the needs of all enterprises concerned. As the food sector builds on food value chains, the difficulties in data management and decision making in food value chains translate into difficulties of the whole food sector in moving forward in new initiatives. It is with this background that the food sector is one of the slowest sectors in adapting new sector encompassing developments including the utilization of modern information and communication technology or the realization of joint initiatives toward improvements in the sustainability of production and trade.

This chapter discusses the issues and provides a framework for supporting data management and decision support in food value chains. Starting from a discussion of the decision situation in food value chains, the chapter outlines a selection of tools for decision support and concludes with a presentation of some decision problems in data management and food chain organization.

## 2 Decisions in supporting data management

### 2.1 Decision situation

In dealing with the complexity of data management and decision situations in food chains, enterprises may face two alternatives in making decisions on activities and developments. They may ask for a decision in issues that:

a) are not dependent on decisions of other enterprises or will not affect other enterprises in the chain (complete decision autonomy); examples concern internal efficiency or enterprise-based quality management; or
b) affect other enterprises, are dependent on other enterprises or are only of relevance, if other enterprises cooperate (limited decision authority); examples include data communication or quality and environmental management across the chain.

The first situation is the traditional enterprise decision situation and will not be dealt with in this chapter. Our focus is on the second situation where any solution has to consider the *interdependencies* within the food chain.

For dealing with such a situation, food chains could agree on a so-called 'chain captain', usually a strong partner in the chain who organizes and coordinates joint decisions. This approach is traditionally favored in situations where a) one partner has a very dominant role such as a regionally dominant dairy company or slaughterhouse and b) partners have close and maybe even exclusive trade relationships. Cases in point were the initial 'IKB chain initiatives' (van Trijp et al., 1997) in the Netherlands where partners in 'almost closed' chains were agreeing on advanced production standards. The drawback of this model is its limited flexibility of aligning agricultural production variations with the need for serving market agreements.

The alternative to following a 'chain captain' is characterized by a network situation where trade relationships along the value chain are dynamically evolving within a food (chain) network situation depending on production and market developments, that is where enterprises change trading partners in response to production and market developments. In this scenario, enterprises are faced with a number of coordination problems including

a) the coordination of agreements on joint action as, for example, on the delivery of products that serve certain quality or environmental needs, and
b) the coordination of agreements on the distribution of costs and gains.

Most of today's discussions on food chain developments focus on the first line of coordination and omit a discussion of the second line. However, the second line of coordination is of crucial relevance for realizing developments that need to find acceptance with enterprises in the long run. It is more the rule than the exception that the costs for realizing certain product characteristics or data collection and processing efforts occur at stages that are different from the stages where the gains are realized. A typical example for engagements in product characteristics involves 'animal welfare' where a higher level of animal welfare is provided through agriculture while higher prices with consumers are realized at retail. The 'animal welfare' example is also representative for an unbalanced situation in data collection and use as most data have to be collected at agriculture, which have to carry the collection costs while the added value of transparency with consumers is realized at retail stage.

In a network situation, the coordination of joint action as well as reaching agreements on the allocation of costs and gains is not easy to reach. The sector usually needs an external facilitator for being able to move forward. The government might take over the facilitator role through regulatory activities. Alternatives are new organizational chain models where partners have agreed on following specific management rules in dealing with coordination needs.

They are usually linked to product or communication standards that provide the base for a common understanding. Product standards may be managed by certain groups of the chains (as demonstrated by the example of GlobalG.A.P., a standard management activity evolving from retail groups; www.globalgap.org), jointly by all groups of the value chain (as demonstrated by the example of Q&S, a standard management group initially evolving from agriculture, www.q-s.de/en/), or by independent standardization organizations (as the example of the ISO standards involving the quality management standards ISO9001, www.iso.org). Communication standards need to build on a) a common technology such as the blockchain technology (Galvez et al., 2018) and b) standards in the infrastructure of communication as provided, for example, by the EPCIS standards of the GS/1 standardization organization (www.gs1.org).

Apart from regulatory initiatives by government, the engagement in standards in the food sector will lead to different levels of adherence, with some enterprises confirming to standards and others not. If companies adhering to different levels of standards engage in trade relationships, the emerging product characteristics will usually be determined by the lower level of standards. A related situation concerns data communication where companies might engage on similar levels of data exchange agreements. If not, the company with lower data exchange engagement determines the data level arriving at consumers' end (Fig. 2).

## 2.2 Decision model approaches

It is obvious that in a chain coordination situation, classical decision models building on mathematical relationships will not work. Decision support is

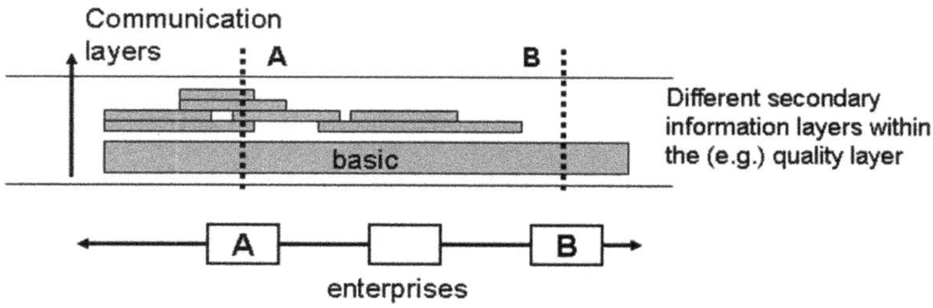


**Figure 2** Quality layer alternatives. Source: author.

needed for a multi-player situation where partners *need to agree* on the various coordination levels, that is, on focus, allocation of activities and distribution of costs and gains. A suitable approach builds on the identification, evaluation and selection of decision alternatives. The following approaches might look simple on first sight, but they are not, as they require substantial knowledge of the complexity of a decision situation and the underlying objectives that reach beyond pre-determined mathematical relationships.

The identification of alternatives might be difficult to achieve but is not dependent on coordination needs. The difficulty is primarily due to the need for specifying boundaries of the decision space. It is obvious that in a long-term view (strategic view) the boundaries are further apart than in a short-term view. The long-term view may consider fundamental changes, while the short-term view is looking more at adaptations of existing solutions to changes in the economic, social and natural environment.

Once the boundaries have been set, there are no 'best' approaches for specifying decision alternatives. It usually builds on more or less structured approaches linking creativity with formal analysis. Long-term views require more creativity, while short-term views might be able to look at eliminating identified deficiencies. For the short-term view various tools have received high visibility, especially in connection with process improvements. They include the 'Failure Mode and Effect Analysis (FMEA)' (Hu-ChenLiu et al., 2013) and the application of the 'HACCP principle' (Hazard and Critical Control Point, www.fao.org/docrep/005/y1390e/y1390e09.htm), which is very much used in, but not restricted to, food safety assurance activities. As an example, the FMEA can be efficiently used in situations where one wants to identify deficiencies in food chain data communication and use and decide about improvements and improvement priorities.

The evaluation of alternatives is crucial. Partners have to agree on common characteristics (usually on multiple characteristics) that could be related to their objectives and would have to be evaluated by each partner regarding his/her individual objectives. As chain companies might arrive at different preferences for alternatives, evaluation needs to be followed by a negotiation process for arriving at a common decision. If the evaluation is based on monetary terms, a joint decision could be easier to reach than if evaluation is based on different characteristics with different scales. Assuming that the monetary advantage of a certain decision alternative is 'best', considering a complete value chain reaching from agriculture to consumers, it would be considered a 'fair' decision if gains are allocated among partners according to engagement in costs.

For supporting the decision process, a number of decision tools have been developed. The availability of different decision tools signals that there is no best tool but that each of the tools considers different aspects. It is therefore common that users will build their decision not just on the results of a single tool but apply different ones and arrive at a decision by evaluating the different 'solutions' selecting one of them, or find a suitable combination. We will present two basic tools, decision tables and the SWOT analysis, which provide insights into principal approaches.

# 3 Decision tools

## 3.1 Decision tables

Decision tables relate decision alternatives to a) future scenarios and b) objectives, considering different possible realizations of objectives due to external influences (scenarios) (Fig. 3). As an example, an objective could represent money return from the realization of an alternative in a certain scenario considering gains and costs. A decision table provides transparency and allows users to identify the alternative that best fits own interests, captured in a 'decision rule'. Literature has developed various decision rules for finding the 'best' alternative (McFadden, 2001). In the example, the rule assumes that the expected objective value (EV) of an alternative and its maximal value across alternatives are used for selecting the best alternative. However, such a rule can provide only suggestions. The final decision is the responsibility of the enterprise that will follow its own 'rule'.

The decision table can be adapted to situations where a multiple of objectives needs to be considered. It is an established procedure to 'translate' each of the objective values related to an alternative and a single scenario into

| Decision alternative | Scenario 1 | Scenario 2 | Scenario 3 | Expected objective value |
|---|---|---|---|---|
| Alternative a | Obj.value a1 | Obj.value a2 | Obj.value a3 | EV (alternative a): EVa |
| Alternative b | Obj.value b1 | Obj.value b2 | Obj.value b3 | EV (alternative b): EVb |
| Alternative c | Obj.value c1 | Obj.value c2 | Obj.value c3 | EV (alternative c): EVc |
| Probability p of scenario occurrence | p1 | p2 | p3 | Rule: max {EVa, EVb, EVc} for selecting alternative |

**Figure 3** Decision table.

a common scale, for example, a scale reaching from 0 to 10. This transfer of an objective value into a standardized scale reaching from very low (0) to very high (10) involves an evaluation of the objective value regarding its positioning within a sensible range.

The scale values of the different objective values are added, considering weights that express their relevance for the company (Fig. 4). This results in a single scale value, related to the alternative in the selected scenario. The further procedure is then similar to a single objective situation building on decision rules.

## 3.2 SWOT analysis

The view of the SWOT analysis differs from the view of the decision table. It involves a dynamic view, starting in:

a) a first step from a present situation and moving in; and
b) a second step to a situation one would encounter after alternatives have been realized.

It is less concrete than the decision table but links its evaluation to the situation of the enterprise (or chain). Each step is captured in a cross table, which lists the strength and weaknesses of the enterprise or chain and the opportunities and threats to the enterprise or chain related to available decision alternatives.

Strengths represent an enterprise's (chain's) success criteria, weaknesses its deficiency criteria. Opportunities and threats are related to decision alternatives that, if realized, could improve the strengths or increase/decrease weaknesses of an enterprise (chain). The comparison of strengths and weaknesses in the two cross tables (present situation and future situation) provides the base for finding a decision. The decision support might be called a 'soft' support as it does not include mathematical relationships or the formulation of results in numeric terms. However, by analyzing the tables, decision makers get an impression of the comparative advantages and disadvantages of different alternatives based on his/her implicit evaluation of relevance (Fig. 5).

It is common to capture the two cross tables in a two-dimensional graph that displays a single value balance of strengths and weaknesses and a single value

**Calculation of a cell scale value** (Alternative X, Scenario Y) in an example for two objectives:
Objective 1 (Alternative X, Scenario Y): objective 1 value --> scale value 1 (value 0–10)
Objective 2 (Alternative X. Scenario Y): objective 2 value --> scale value 2 (value 0–10)
**Summary cell scale value** (Alternative X, Scenario Y): w*scale value 1 + (1–w)*scale value 2

**Figure 4** Calculation scheme for a cell in the decision table for two objectives.

| Opportunities for enterprise (chain) | Strengths of enterprise (chain) |
|---|---|
| > List of opportunities<br>.......... | > List of strengths<br>.......... |
| **Threats to enterprise (chain)** | **Weaknesses of enterprise (chain)** |
| > List of risks (threats)<br>.......... | > List of weaknesses<br>.......... |

**Figure 5** Cross table of SWOT analysis.

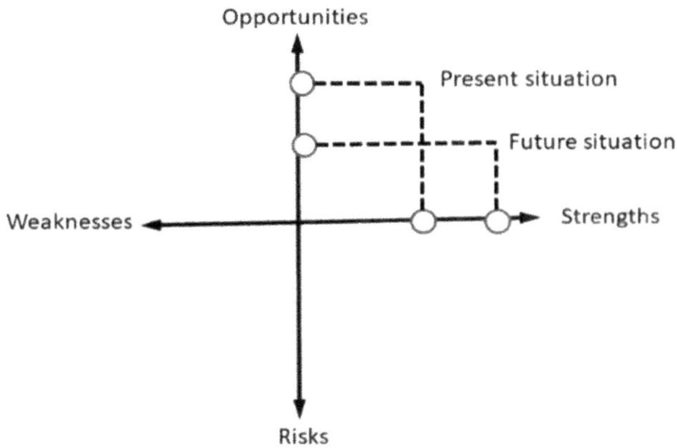


**Figure 6** SWOT analysis graph. Source: author.

balance of opportunities and threats for the present and the future situation, both translated into a basic scale. Such balances are based on decision makers' (rough) estimates of relationships and weights (Fig. 6).

In the graph, the future situation (after the implementation of the decision alternative) is characterized by a lower level of opportunities (as one opportunity has been realized already) and a higher level of strengths due to the realization of the alternative. If different alternatives are up for decision, the consequences of the alternatives can be captured in different SWOT tables and integrated into the graph. The graph could be a base for further discussions. As an example, if two different decision alternatives would lead to two different future situations, the future situation with the highest contribution to an enterprise's (or chain's) strength might look most attractive. However, if it is linked to a lower level of remaining opportunities, an enterprise might lose flexibility for future activities. This asks for a detailed analysis of future needs and expectations.

Suitable decision alternatives identified with a SWOT analysis can be integrated into a decision table for reaching a final decision building on a more concrete specification of objective values, for example, in monetary terms.

### 3.3 Multi-criteria decision tools

With many different characteristics to consider in deciding about decision alternatives, a variety of additional supporting approaches summarized as 'multi-criteria decision analysis' have been developed. Most of them are beyond the discussion in this chapter. They all involve the need for certain mathematical calculations of varying complexity. However, for integrating different decision makers into a joint decision process with a number of decision alternatives, the analytical hierarchy process (AHP) has gained some prominence (Saaty and Vargas, 2001). It is basically focusing on supporting the negotiation process through a structured process in which the different decision makers engage in pairwise comparisons of different alternatives. The difficulty is in the decomposition of the decision situation into a number of sub-decisions that can be dealt with independently. The AHP involves a complex mathematical model that summarizes the results of the sub-decisions into an overall 'group decision'. This outline demonstrates the complexity of the approach, which usually is supported by computer-based model calculations.

## 4 Principal case studies

### 4.1 Data management for transparency

A specific decision situation concerns data exchange between enterprises along the chain and toward consumers. It is a core feature for realizing transparency and trust, which is gaining increasing attention by consumers and which depends on the availability of suitable data and their consideration in suitable decision activities. We will not elaborate on problems that may arise because of data manipulations in sourcing and/or along the chain. There is a lot of discussion going on. The relevance of both issues can be reduced through technological developments. Data sourcing can be improved through emerging sensor technologies (Grossi et al., 2019) and data communication through blockchain concepts (Galvez et al., 2018).

The remaining problems concern agreements on data collection and data exchange, which require coordination among participants. The coordination problems resemble the general coordination problems in production issues. Enterprises have to agree on what data to collect, where to collect, when to collect, what to communicate, how to communicate regarding format and timing and where to use, by whom and what for. The relevance of the last issue, the clarification of what data are being used for, has frequently been underestimated. There are examples especially from quality management where data collection in agriculture failed because farmers did not trust retailers' communicated interest and where afraid, data could be used at their disadvantage. In a network situation, there might be different levels of agreements, which at the end will provide different levels of data for retail and its ability to inform consumers.

One of the major difficulties in improving transparency and trust is due to the fact that most information that might be relevant for consumers is to be collected in agriculture while retail is the main benefiter. Examples presently in discussions concern the use of pesticides in plant products, especially vegetables and fruits and the consideration of animal welfare issues. These differences are one of the major barriers in improving data communication along chains as agriculture is not sufficiently benefiting from its data collection and the related costs. Furthermore, in present organizational schemes for data exchange, agriculture is no longer considered owner of data as soon as they have left its premises. As a consequence, agriculture has no longer influence on what its data are being used for, a major obstacle for improvements in data communication.

That is a situation where new data collection and communication alternatives have to be identified and evaluated by all concerned. Sensor-based technology might help cutting costs for data collection and transfer. An appropriate allocation of costs, benefits and ownership asks for more radical new opportunities. One of the opportunities discussed in literature that could provide support in solving the cost and benefit problem is the establishment of 'data markets'. Agriculture would offer data (e.g. through an online platform), and trading partners could make offers for receiving data. This alternative could easily deal with differences in data intensity. Such approaches could lead to differences in product offers in retail shops, products with less and products with more information on issues of interest to consumers and in consequence, with differences in prices. Groups with already-intensive data collection initiatives in agriculture such as GlobalG.A.P. have for long discussed the need for differing products in retail subject to the availability of product information.

A simplified version of a data market has been established in the 'book-and-claim' concept that has been used in soybean production and sales (https://greenpalm.org/about-greenpalm/why-greenpalm-makes-a-difference/book-and-claim-supply-chain-model). In the production of high-quality soybeans, the quality information captured in a quality label (representing a data cluster) is being separated from the actual products. Products and labels are then marketed separately. The products are channeled into the regular soybean market without any link to the initial label information (Fig. 7).

At the end of the chain, buyers of soybean could purchase a fitting number of labels and attach them to the products, selling them as 'high-quality' products. This approach of separating data from products works well, if data refer to quality issues that are not linked to measurable product characteristics but refer to credence attributes such as fair trade or environmental-conscious production. It assures that the sales quantity of 'high-quality products' matches

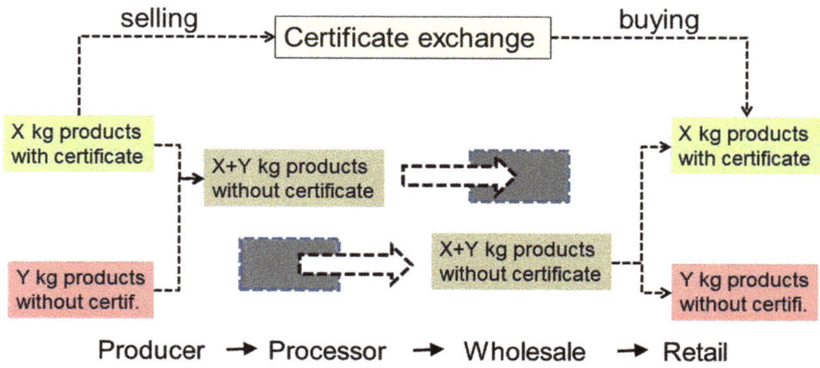


**Figure 7** Book-and-claim approach. Source: author.

the production quantity. In cases where data refer to measurable product characteristics, data markets have to assure that traded data remain linked to the specific products they refer to.

In data management, enterprises are, in addition, confronted with problems that reach beyond data coordination problems. They refer to the differences in enterprises' ability to collect, communicate and use data, known as e-readiness (Bryceson, 2006). While one enterprise might be able to rely on a sophisticated digital data processing and communication ability, another one might still rely on paper-based data management. FMEA could help identify areas of weakness. The level of data communication along the chain is determined by the lowest level of e-readiness in any of the participating enterprises. In making improvements to communications, it may be necessary to agree the costs and resources required to bring all enterprises up to the level of efficiency, transparency and trust all members of the chain require.

## 4.2 Quality and environmental management

Quality and environmental management activities play an important role in agriculture and the food sector. Quality management assures the delivery of food to consumers, which is safe to eat and of the quality that trading partners and consumers expect. Environmental management serves the expectations of society in the protection of the environment and in an environmentally sustainable production.

The need for the coordination of activities along the food chain is quite similar in quality and environmental management. The coordination need is driven by regulatory requirements of public authorities and by pressures of markets. In consequence, the need to cooperate for coordination is much higher than in data coordination. Regulatory authorities have since many years and especially since the bovine spongiform encephalopathy (BSE) crises in

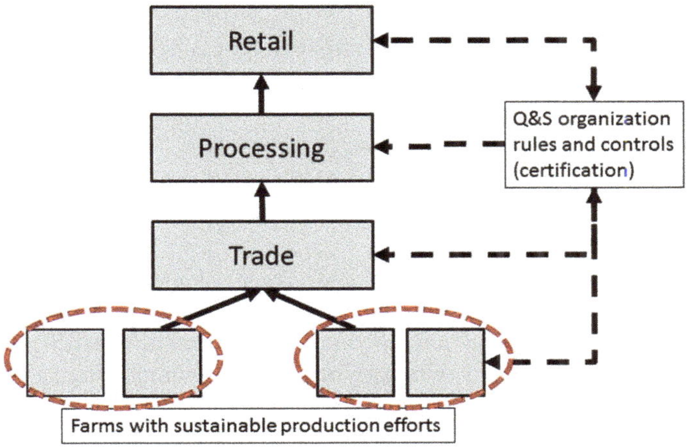


**Figure 8** A basic quality coordination scheme. Source: author.

Europe enforced coordination activities not only regarding the assurance of food safety but also regarding the avoidance of environmentally damaging activities (Fig. 8).

As a result, there are a number of coordination models that are being tried out in real-life 'experimental settings'. They all build on a continuous coordination model, as requirements of policy and markets on food quality and environmental protection are continuously changing. All of them have advantages and disadvantages, which qualify them as decision alternatives, companies and food chains could select from. Differences concern the structural organization, the level of information exchange, the tackling of problems dealing with differences in the allocation of costs and benefits along the chain and the engagement of monitoring and auditing schemes for assuring companies' adherence to the coordination model agreed upon.

It is important to note that all alternatives do not build on a coordinated data exchange along stages of food chains throughout the sector, which allowed for the dynamics in food chain trade relationships but find opportunities to 'simulate' data exchange through a variety of means. A coordinated data exchange throughout the sector that could serve the needs of all stakeholders concerned are still far into the future.

A basic approach is realized through external quality management audit schemes as exemplified by the schemes Q&S in Germany (https://www.q-s.de/en/) and IKB in the Netherlands (http://www.prnp.nl/public/doc/PRLT_4230_IKBdiervoeder.pdf). The coordination is not between individual enterprises but through an external organization. It specifies production requirements for each individual stage of the food value chain, which assures certain agreed levels of food quality and environmental protection. With this approach, companies do not have to exchange data on food quality or environmental issues as they 'know' what requirements have been fulfilled by its suppliers. In addition, they are not bound to a certain supplier but could choose between all suppliers that follow the scheme. This simplifies the implementation of a coordinated quality and environmental quality management scheme not only along a chain but within the whole food sector. It is obvious that such a scheme depends on a reliable monitoring and audit activity that assures the adherence of all companies to the scheme.

However, the scheme also has drawbacks. First, it is not considering coordination in the allocation of costs and benefits. Furthermore, as its focus is on assuring certain levels in quality and environmental protection across a whole sector, improvements in quality and environmental protection have to assure that the majority of enterprises are prepared to follow. The realization of this decision alternative is, therefore, most suitable for situations where a sector-wide basic level of quality assurance and environmental protection needs to be implemented.

At the other end of coordination intensity is a scheme that tries to eliminate the drawbacks but requires, on the other hand, a coordinated data collection and exchange for reaching higher levels of quality assurance and environmental protection, which supports the visibility of products in the markets and a clearer identification and separation of additional costs and benefits.

In this scheme implemented by a major retailer, the final consumer products must fulfill higher levels of quality assurance and environmental protection based on appropriate requirements on all stages of the chain. These higher levels will be communicated to consumers through specific brand names that will allow higher prices. The additional gains are partly channeled back to agriculture, which carries the major burden in reaching the quality and environmental requirements (Fig. 9).

In an implementation of this scheme in the market, the scheme involves the consideration of high levels of environmental issues that are captured in an index that allows some flexibility in production. The final products must reach a certain index level to be accepted, carrying the exclusive brand name. It is obvious that the formulation of such an index requires intensive data collection and data use at each stage of the food chain. However, as in the preceding example, the assurance of quality and environmental protection builds on the establishment of an intensive monitoring and audit scheme and not on the information exchange between enterprises. This pays tribute to the data coordination difficulties discussed earlier and the still poorly developed schemes for assuring reliability in data collection, processing and communication. New developments in data collection and communication technology including sensors, blockchain technology, smart contracts linked

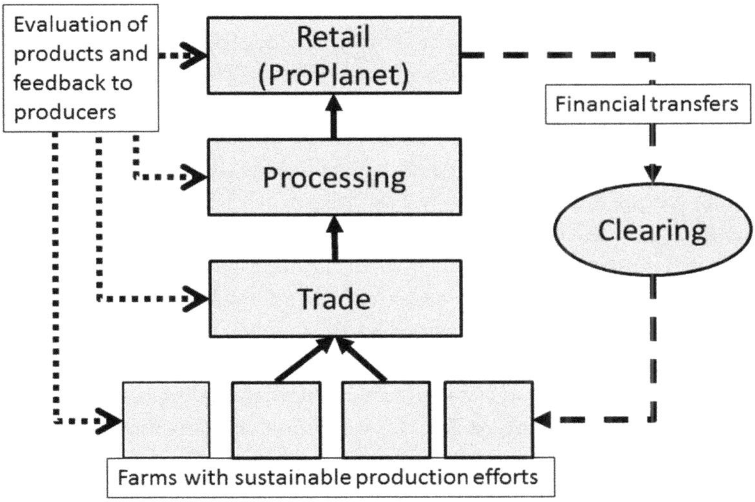


**Figure 9** An advanced coordination scheme.

to blockchain technology and others will change the picture in the future and reduce the need for external monitoring and auditing activities.

# 5 Conclusion and future trends

## *5.1 Summary*

Decision support in food chains is a complex issue as it needs to balance the interests of different actors with different objectives and interests. An approach used sometimes in literature involves the use of chain optimization models. However, these models were not well suited for arriving at solutions that capture the complexity and differences in the interests of the different participants in the chain. In the past, agriculture was very much dependent on decisions of processing and retail companies. With the need for an increase in food production without a similar increase in available resources, agriculture is improving its negotiation position in the chain. Decision making that considers the various partners in the chain as being of similar relevance needs to build on a number of coordination and negotiation activities. This is a difficult process that, however, can:

a) be supported by a number of easy-to-use decision tools, and
b) build on chain organization models that are being tried out in various implementations.

The chapter introduces two alternatives of decision tools that are well suited for considering the complexity in food chain coordination, decision tables and the SWOT analysis. Both tools allow within a rigid structural approach the consideration of individual opinions of food chain participants toward reaching a common decision outcome. However, decision activities require the availability of appropriate data and, in decisions with chain relevance, data from other stages of the food chain. The collection and communication of data along the chain is dealt with through varieties of organizational concepts. As of today, a seamless coordinated flow of all necessary data across the food chain is still limited to specific cases with fixed trading relationships, while the food sector is generally comprised of a network of enterprises where food chain relationships are developing dynamically through changing trade relationships. A coordinated data communication across the whole food sector is still a challenge for the future. The organizational relationships developed for quality and environmental management offer suitable alternatives.

## 5.2 Future trends and information support

The sector is continuously challenged to improve in serving changing needs of consumers while following sustainability and environmental expectations of society. The dynamics in changing consumer needs regarding products and production processes dealing with, for example, animal welfare, avoidance of pesticides, fair trade and related issues requires a continuous adaptation of decisions and data needs across the sector. These challenges cannot be solved by individual enterprises. The same is true for food chains that, as was discussed earlier, are not a stable construct but a dynamically evolving organizational scheme depending on actual trading activities. The challenge is for the sector as a whole that requires sector-wide coordination initiatives and the engagement of institutions who could act as facilitator or guidance by policy. Technological developments have reached a stage where further improvements in data management and decision coordination across food chains are no longer a technological but an institutional problem where the sector has to find suitable means for coordination and the balance of interests.

# 6 References

Bryceson, K. P. 2006. *E-Issues for Agribusiness: The What How and Why*. CABI Publishing, Oxfordshire.

Galvez, J. F., Mejuto, J. C. and Simal-Gandara, J. 2018. Future challenges on the use of blockchain for food traceability analysis. *Trends in Analytical Chemistry* 107, 222–32. doi:10.1016/j.trac.2018.08.011.

Grossi, M., Berardinelli, A., Sazonov, E., Beccaro, W. and Omaña, M. 2019. Sensors and embedded systems in agriculture and food analysis. *Journal of Sensors*. Special Issue 2019, 1-2. doi:10.1155/2019/6808674.

Liu, H., Liu, L. and Liu, N. 2013. Risk evaluation approaches in failure mode and effects analysis: a literature review. *Expert Systems with Applications* 40(2), 828-38. doi:10.1016/j.eswa.2012.08.010.

McFadden, D. 2001. Economic choices. *American Economic Review* 91(3), 351-78. doi:10.1257/aer.91.3.351.

Saaty, T. L. and Vargas, L. G. 2001. *Models, Methods, Concepts and Applications of the Analytical Hierarchy Process*. Kluwer Academic Publishers, Boston.

van Trijp, H. C. M., Steenkamp, J.-B. E. M. and Candel, M. K. J. M. 1997. Quality labeling as instrument to create product equity: the case of IKB in the Netherlands. In: Wierenga, B., van Tilburg, A., Grunert, K., Steenkamp, J.-B. E. M. and Wedel, M. (Eds), *Agricultural Marketing and Consumer Behavior in a Changing World*. Springer, Boston, pp. 201-15.

www.ingramcontent.com/pod-product-compliance
Ingram Content Group UK Ltd.
Pitfield, Milton Keynes, MK11 3LW, UK
UKHW021843100926
8556IPUK00019B/1025

9 781801 46665